Seismic Risk Assessment and Design of Building Structures

INEEC Series on Engineering Aspects of Earthquake Phenomena

Seismic Risk Assessment and Design of Building Structures

Edited by

A. Koridze

Secretary of UNESCO International
Network of Earthquake Engineering Centres

OMEGA SCIENTIFIC

Published in the United Kingdom by
Omega Scientific
John Wilder Building
Hithercroft Industrial Estate
Wallingford
Oxon OX10 9AR
England

British Library Cataloguing in Publication Data

Seismic risk assessment and design of
building structures.
1. Earthquake engineering
I. Koridze, A.
624.1′762

ISBN 1-870199-04-9

Printed and Bound in Great Britain by
Hartnolls Limited, Bodmin, Cornwall.

Preface

Since the publication about a year ago of the first volume of *Engineering Aspects of Earthquake Phenomena* within the framework of activities promoted by the UNESCO International Network of Earthquake Engineering Centres (INEEC), a number of seminars, workshops and training courses in various fields of earthquake engineering have been organized or supported in Italy, Yugoslavia, Turkey, Morocco, Tunisia, Venezuela and the USSR.

The last international seminar was organized by UNESCO, in cooperation with the Academy of Sciences of the Georgian SSR, in Tbilisi, USSR, in October 1987. This seminar was combined with the inauguration of the UNESCO–INEEC International Centre for Earthquake Engineering in Tbilisi, which is responsible for the development and promotion of seismic codes and standards in developing countries. The objectives of the seminar were to review the state of the art of engineering seismology, the aseismic design of various engineering structures, including school buildings, and the improvement and development of national seismic codes and standards. The papers published here were selected from those presented at the seminar; several contributions from eminent scientists that were recommended for publication by the review panel of the INEEC Scientific Council are also included.

In publishing this second volume in the Engineering Aspects of Earthquake Phenomena series, UNESCO INEEC intend to continue a series of annual publications which will not only serve for the dissemination of new ideas in earthquake engineering but also help in the transfer of knowledge from the world centres of excellence to concerned centres in developing countries.

A. Koridze, UNESCO

Contents

Earthquake Vulnerability and Loss Assessment for Physical and Urban Planning

JAKIM PETROVSKI

Institute of Earthquake Engineering and Engineering Seismology (IZIIS), University "Kiril i Metodij", Skopje, Yugoslavia

Abstract Earthquakes have become increasingly destructive in recent years, particularly in Mediterranean countries, because of the growth in population and increasing urbanization. The cost of replacing and repairing earthquake-damaged buildings is a significant drain on the economies of earthquake-prone countries. Therefore it is imperative to achieve methods of reducing earthquake damage to an economically supportable level. To this end we have developed a method of assessing risk to buildings on the basis of vulnerability functions. Empirical vulnerability functions were derived from data obtained from communes in Montenegro, and theoretical vulnerability functions were also calculated. The results are expressed in graphical form.

1 *Introduction*

Over the last two decades natural disasters, particularly earthquakes, have become increasingly destructive as they affect larger concentrations of population and property. Orderly industrial development accompanied by urban expansion and increased population is prohibitively expensive in earthquake-prone regions unless investments in infrastructure, housing and other public and social facilities are protected against damage at all stages of their development. Although significant efforts have been made to assess seismic hazards and mitigate their possible consequences, major earthquakes have continued to cause enormous damage to the economy of the regions directly concerned and the country as a whole. For example, the direct economic losses caused by the Montenegro earthquake of 15 April 1979 alone were estimated at about 10% of the gross national product of Yugoslavia for 1979. A similar event, the 1963 Skopje earthquake, cost the Yugoslav economy about 15% of the gross national product for 1963. When other events that have taken place in Yugoslavia between 1963 and 1984 are taken into account (Petrovac, 1966; Debar, 1967; Ulcinj, 1968; Banja Luka, 1969 and 1981; Kopaonik, 1980, 1982, 1983 and 1984), the penalty paid for earthquakes is estimated at an average of more than 1.5% of the gross national product per annum.

Developments in earthquake engineering over the years have enabled many countries exposed to high seismic hazard to decrease the overall seismic risk to populated regions by relocating highly vulnerable structural systems according to new policies in urban planning or by replacing conventional structural systems with modern systems designed and built according to recent seismic regulations and codes. However, a large number of traditional structures which are not able to resist earthquakes without significant damage and collapse still exist. This makes the overall process of improvement too slow, and earthquakes will continue to cause seismic disaster manifested in terms of damaged or destroyed structures and lifeline systems, loss of human lives, and other direct and indirect losses including long- and short-term adverse effects on the economy of both the region and the whole country. Therefore the principal goal of engineers, urban planners, public policy makers and administrators is to achieve a rational strategy for earthquake safety programmes, i.e. to protect earthquake-prone regions or to enable the effective rehabilitation of economic and social activities in earthquake-stricken regions.

2 Assessment of expected vulnerability and seismic losses

An integral model for the assessment of expected vulnerability and seismic risk in the region under consideration should include the following basic steps:

(1) identification of the elements at risk;

(2) evaluation of the seismic hazard;

(3) derivation of appropriate vulnerability functions for given elements at risk, which describe the interrelation between the specific loss and the seismic hazard;

(4) evaluation of the specific seismic loss per element at risk and its contribution to the existing total risk;

(5) evaluation of the total seismic risk.

Of the three factors determining seismic risk, i.e. the number of elements at risk, their vulnerability (specific loss potential) and the seismic hazard, only the first two can be controlled during physical and urban planning or through pre-disaster prevention programmes. Although it is possible to control the value of elements at risk efficiently by relocating those which are sensitive to seismic exposure in regions of lower seismicity, it is still necessary to provide economically justified practical measures for protecting other elements which, owing to favourable economic and environmental conditions, have to be located in regions of higher seismicity. In the latter case a level of acceptable risk depending on the economic development of the region or the country should be estimated and defined, and the level of vulnerability should be modified by improving the seismic resistance of the elements at risk or decreasing the density of population and man-made structures.

3 *Identification of elements at risk*

Since an earthquake is a precipitate phenomenon which is outside human control and which causes economic loss, it is necessary to identify the exposed elements at risk, i.e. to define a loss model for the region.

A direct loss model usually describes physical damage expressed in terms of human casualties and injuries, damage to local infrastructure (road and railway systems, water and gas supply etc.), damage to residential and other buildings, and damage to any material goods. In addition to the physical damage and functional disruption caused by earthquakes, there are also categories of indirect effects which can generally be classified as economic and social damage. Stagnation of industrial activities, a decrease in industrial production and regional revenue, and extra expenditure for immediate rehabilitation of the stricken area are classes of typical indirect economic loss. Interruption of transport, water and electric power supply systems, decrease in public and information services and adverse effects on the reputation of the damaged areas are examples of social damage.

In order to estimate correctly the magnitude of a possible disaster and the level of capital investment that should be allocated in advance for the mitigation of the consequences of an earthquake, it is necessary to formulate a model for assessing the long-term effects of earthquake losses. This can be done efficiently by coupling the direct loss model and the indirect effects model through interrelation and performing a time-series estimation of the economic effects induced by direct losses. It is also necessary to have a mechanism for adjusting between the unaccelerated productivity due to earthquake-induced damage and the accelerated demand of investment for repair and reconstruction of the stricken region.

4 *Determination of seismic hazard*

Seismic hazard can be determined for different return periods with respect to the economic lifetime of the elements at risk. Vibrational effects and geological hazards in densely populated areas should be determined by carrying out seismic microzoning studies.

The most frequently used method for seismic hazard analysis is that of Esteva (1964) and Cornell (1968, 1971) which was developed to produce relationships between ground motion parameters such as the modified Mercalli intensity, the peak ground acceleration, the peak ground velocity etc. and their average return period for a given site taking into account arbitrary geographical relationships

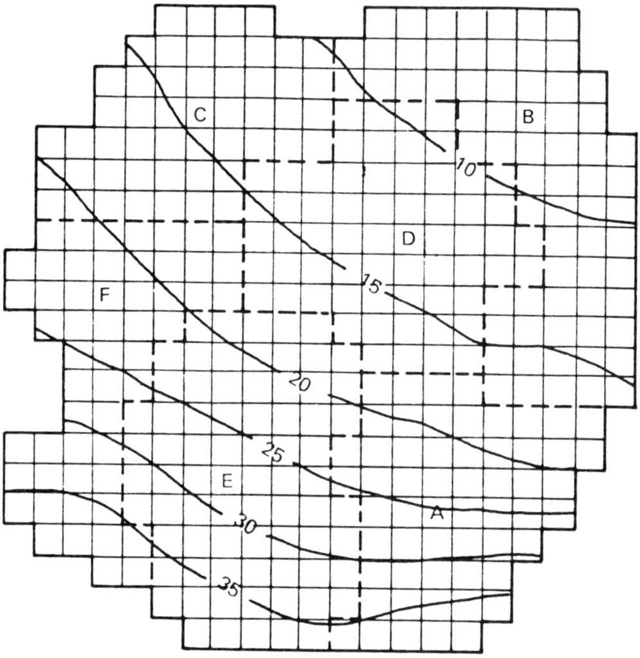

Fig. 1 *Seismic hazard map for a 200 year return period (SH-200).*

between the site and the potential point, line or area source. The Cornell–Esteva technique provides a method for integrating the individual influences of potential earthquake sources (local or distant, strongly active or weakly active) into the probability distribution of maximum annual quantities (intensity, peak ground acceleration or any other quantity defined by the user). The results from a grid of different sites in a region can be used to construct seismic hazard maps which show, for example, contours of equal average peak ground acceleration with a given probability of not being exceeded for a specified time period (Fig. 1) or contours of equal average return period for certain levels of maximum peak ground acceleration (Davenport and Milne, 1969). The results of the development appear in closed analytical form, permitting direct observation of the sensitivity of the final results to the estimates to be made.

5 Earthquake damage assessment

The development of an earthquake loss model including the various elements at risk and the assessment of seismic risk for a region are strictly dependent on damage data from past earthquakes. Since a significant increase in damage is expected from future major earthquakes in a large number of countries in seismically active regions of the

Mediterranean, the development of a unified methodology and procedure for examining and reporting earthquake damage to buildings and other structures in urban and rural areas is essential for establishing a database of earthquake effects and a procedure for assessing earthquake damage and economic losses. The primary purpose of these procedures and methodology is to ensure that there are sufficient data for the following needs of the community and national authorities:

(1) reducing the incidence of death and injury to occupants of buildings which have been weakened or seriously damaged by seismic activity and which will probably be subjected to series of aftershocks for several months after the principal shock;

(2) obtaining appropriate information on the magnitude of the disaster in terms of the number of usable, damaged and dangerous buildings for the purposes of the immediate protection of human lives, housing the populace and enabling fundamental human activities to be carried on in the affected region;

(3) providing a database for the estimation of economic losses, for developing an appropriate rehabilitation programme and for assisting the reconstruction and development of the affected region;

(4) creating a database for the prediction of the consequences of future earthquakes and the behaviour of other seismic regions;

(5) planning and organizing relief including rescue operations after an earthquake, staff training and organization of supplies;

(6) recording and classifying damage so that earthquake-damaged buildings can be repaired and strengthened;

(7) identifying the principal elements of the earthquake loss model and developing vulnerability relationships for various categories of buildings so that short- and long-term actions for reducing the consequences of earthquakes can be established and pre-earthquake assessments can be performed;

(8) improving seismic design and construction codes, regulations and practice;

(9) improving the scientific base for physical and urban planning and for general planning for the reduction of the consequences of earthquakes and the mitigation of seismic risk in seismically active regions.

Post-earthquake damage evaluation should be organized by implementing a systematic methodology and a rapid procedure for providing basic information to the local and national government authorities for decision making and for undertaking economically justified and technically consistent measures for reducing the consequences of earthquakes in a uniform manner over the whole country. On the basis of a uniform methodology and procedure and systematically coordinated efforts, it should be possible to develop more practical and transferable data which are required for more efficient seismic risk reduction planning in countries located in seismically active regions of the world.

6 Derivation of empirical and theoretical vulnerability functions

In order to forecast the damage which is expected to occur during future earthquakes, it is necessary to know how various types of structure have behaved when exposed to tremors of different intensities. This information is also essential for cost–benefit studies to determine the relative effectiveness (decrease in specific loss) of various measures which can be used to mitigate seismic risk. Buildings of similar type and size do not necessarily respond in the same way to tremors of a given intensity. Therefore damage must be documented for a sufficient number of similar structural types for the same shaking intensity to enable an average damage level and dispersion estimators of damage to be determined and allocated to various classes of building.

An inventory and classification of damage, usability and serviceability of structures carried out on the territory of seven Montenegro communes for the existing stock of facilities, engineered structures, buildings, houses, utilities etc. was directed by the Republic and District Commissions for Classification of Buildings Damaged by Earthquakes on the Territory of SR Montenegro. It was performed by 500 specialists from the whole of Yugoslavia organized in about 100 working groups and 20 teams of specialists for classifying damage to industrial structures, port and harbour facilities, regional and local infrastructural networks and other structures of vital importance.

Out of the total of over 60 000 buildings classified, the data obtained on over 40 000 structures of gross area 6 715 169 m^2 damaged by the Montenegro earthquake of 15 April 1979 were statistically processed in order to derive empirical vulnerability functions. The damage data used in this study are based on a damage inventory performed for about 334 settlements located in the territory of the six coastal communes of Ulcinj, Bar, Budva, Kotor, Tivat and Herceg Novi and the inland commune of Cetinje. The empirical vulnerability functions were derived using correlations established between the damage levels and the earthquake ground motions recorded on 29 strong motion instruments (SMA-1) and 21 seismoscopes. Generalized physical vulnerability functions and data scattergrams for 21 873 stone masonry buildings classified into three basic categories of damage and usability are presented in Fig. 2. Site-dependent and generalized vulnerability functions derived for four basic structural types of traditional and modern constructions are presented in Fig. 3. It is evident that the threshold of the damage level is quite low (about 5% for 0.1g peak ground acceleration) for the aseismic structures (strengthened masonry and reinforced concrete frame structures designed to be earthquake resistant) and unexpectedly high for stone masonry and brick masonry buildings. The slope of the curves is almost equal for all structural types with an increase in peak ground acceleration, approaching an unacceptable damage level for stone masonry and brick masonry buildings.

A detailed response analysis of 104 buildings of different structural types for five expected levels of peak ground acceleration using the criteria of strength (shear base coefficient) and deformability capacity (interstorey drift) has been used to derive

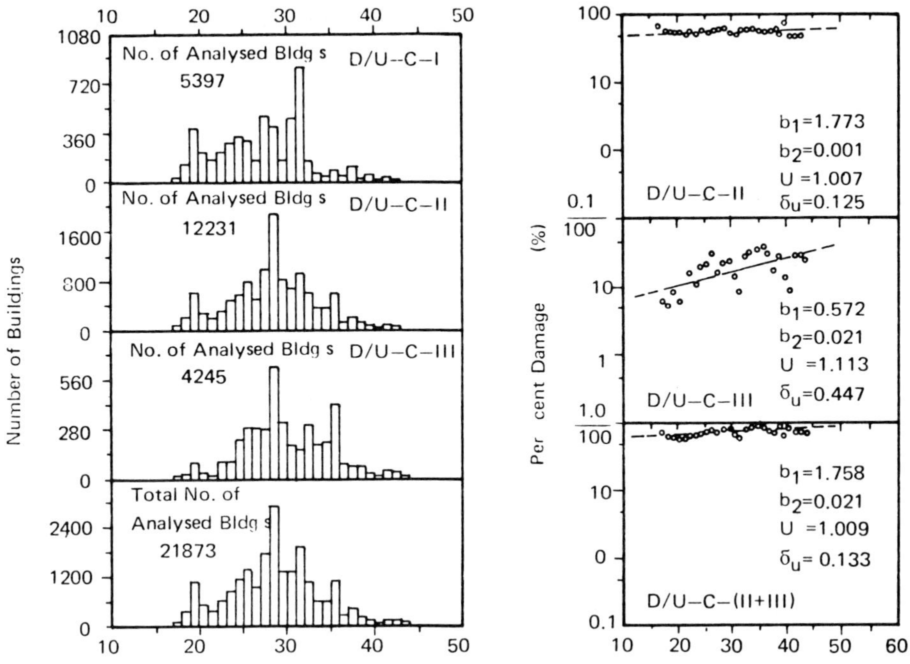

(a) (b)

Fig. 2 Generalized physical (functional) vulnerability functions and data scattergrams for stone masonry buildings: (a) building population analysed; (b) data scattergram and regressed vulnerability functions; PGA, peak ground acceleration; D/U-C, damage/usability category.

theoretical vulnerability functions for various types of aseismic structure (Fig. 4). A detailed cost estimate analysis for repair and strengthening of the same 104 buildings was performed and the results are presented in Fig. 5 in the form of cost functions for the repair and strengthening of unit floor area.

7 Evaluation of specific loss

If a uniform distribution is assumed for each element at risk (by structural type), if empirical and theoretical vulnerability functions are derived and if seismic hazard maps are determined, specific loss mapping can be performed for various categories of damage and structural type as shown in Figs 6 and 7.

8 Cumulative loss—specific risk level

Cumulative loss contours for existing and probable improved scenarios of the

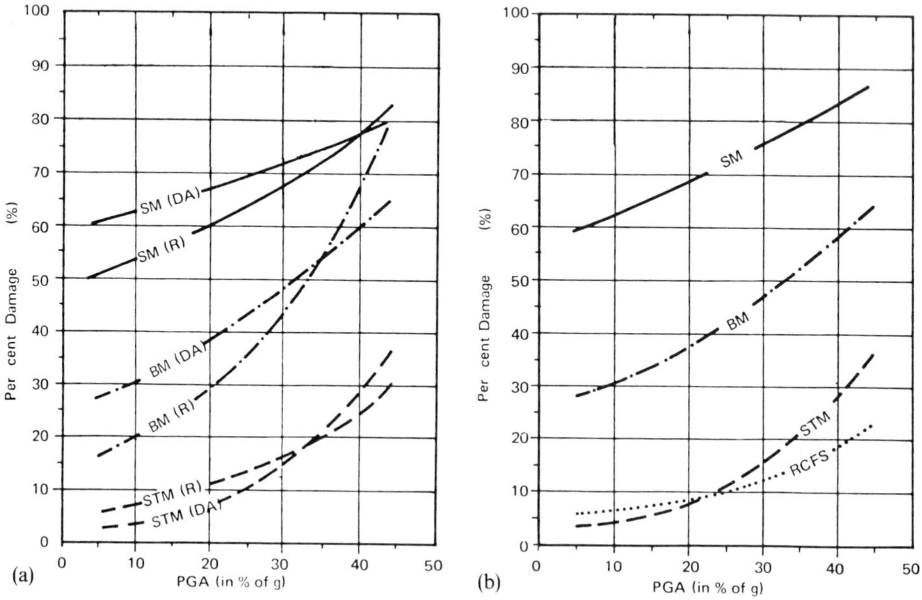

Fig. 3 *Vulnerability functions for damage/usability categories II and III: (a) site-dependent functions; (b) generalized functions; SM, stone masonry; BM, brick masonry; STM, strengthened masonry; RCFS, reinforced concrete frame buildings; DA, diluvial–alluvial soil conditions; R, rock soil conditions.*

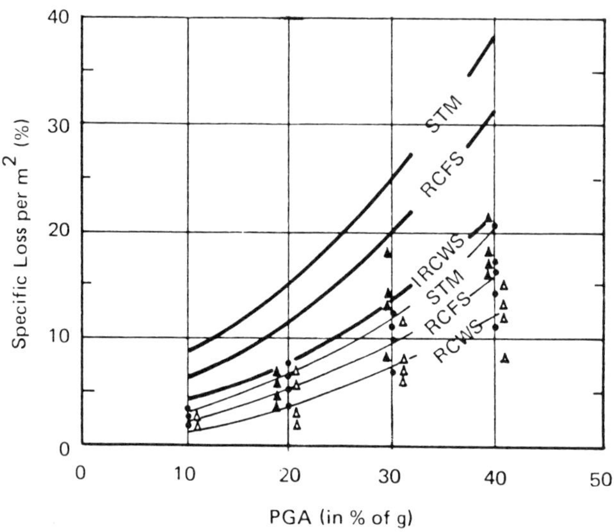

Fig. 4 *Vulnerability functions for the structural systems used in modern aseismic buildings: ▲, STM, strengthened masonry; ●, RCFS, reinforced concrete frame; △, RCWS, reinforced concrete with shear walls; —, total loss; —, structural system loss.*

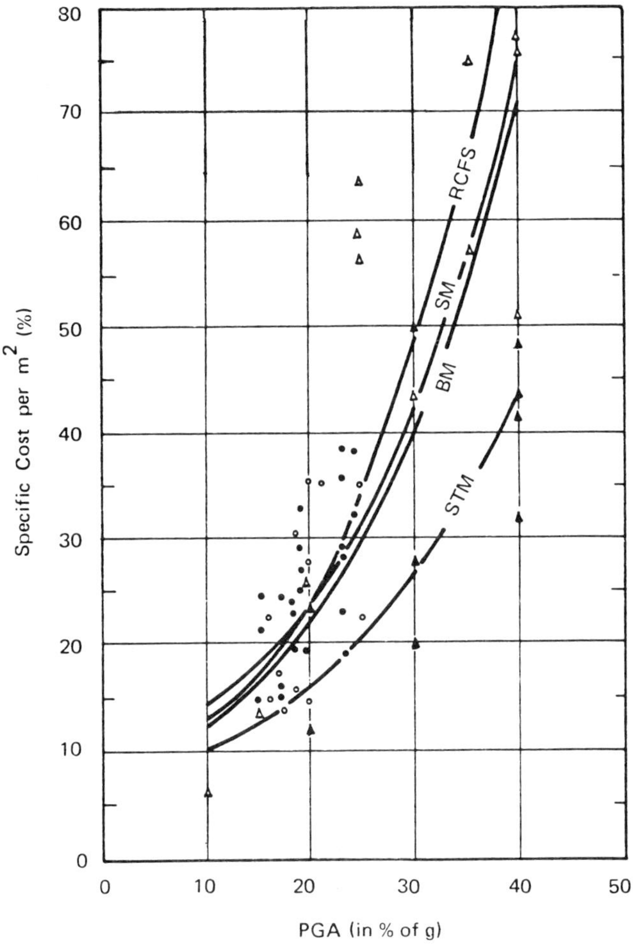

Fig. 5 *Cost functions for the repair and strengthening of various types of building: ●, SM, stone masonry; ○, BM, brick masonry; ▲, STM, strengthened masonry; △, RCFS, reinforced concrete frame.*

presence and density distribution of various buildings at risk (Table 1) are mapped in Figs 8 and 9. Physical, functional and economic loss for each scenario can be assessed on the basis of these maps.

Then either the expected cumulative loss or the "acceptable seismic risk level", whichever is most appropriate, can be determined on the basis of the period of time likely to elapse before the next earthquake and the economic potential, and it can be used in the development of a programme of preventive measures. Alternatively a suitable disaster preparedness programme can be organized to cope with the effects of the expected earthquake.

The cumulative losses of the two scenarios considered in Table 1 and Figs 8 and

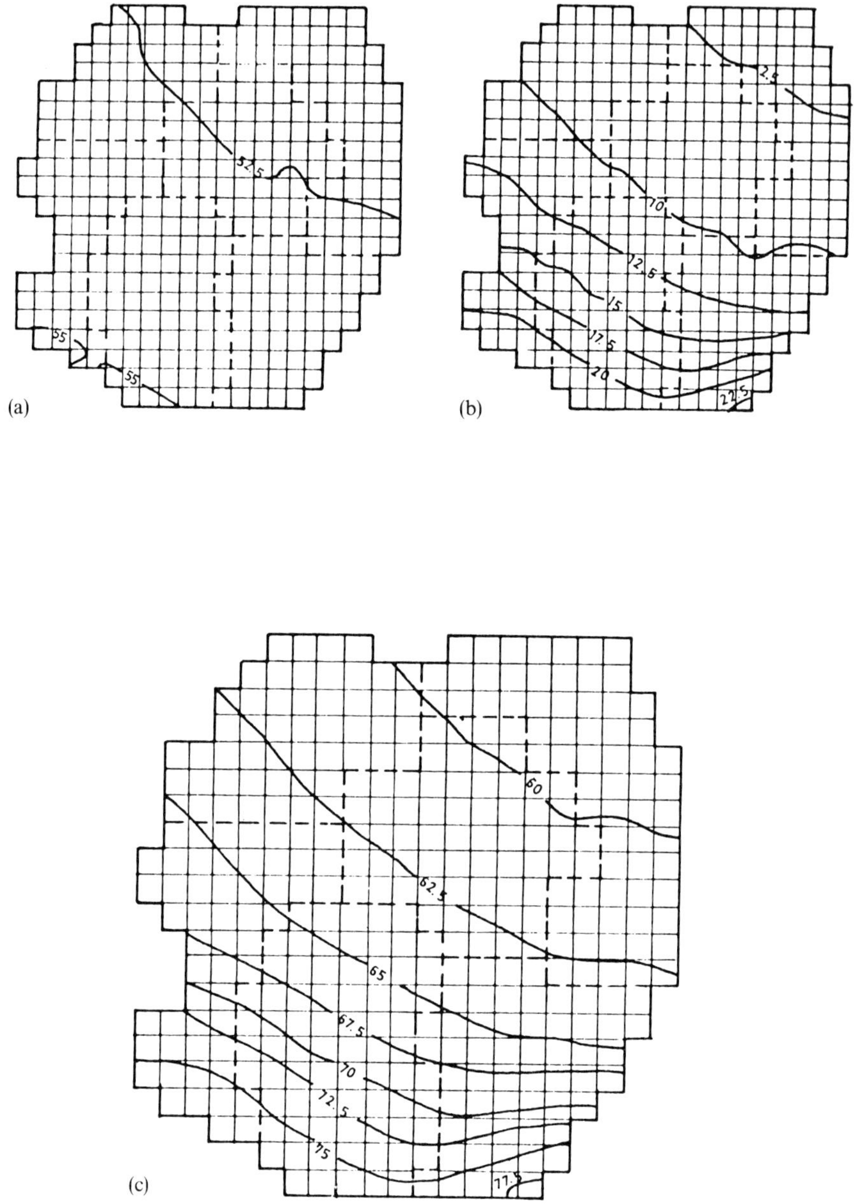

Fig. 6 *Specific loss contours for stone masonry buildings for an earthquake with a 200 year return period: (a) damage/usability category II; (b) damage/usability category III; (c) total for damage/usability categories II and III.*

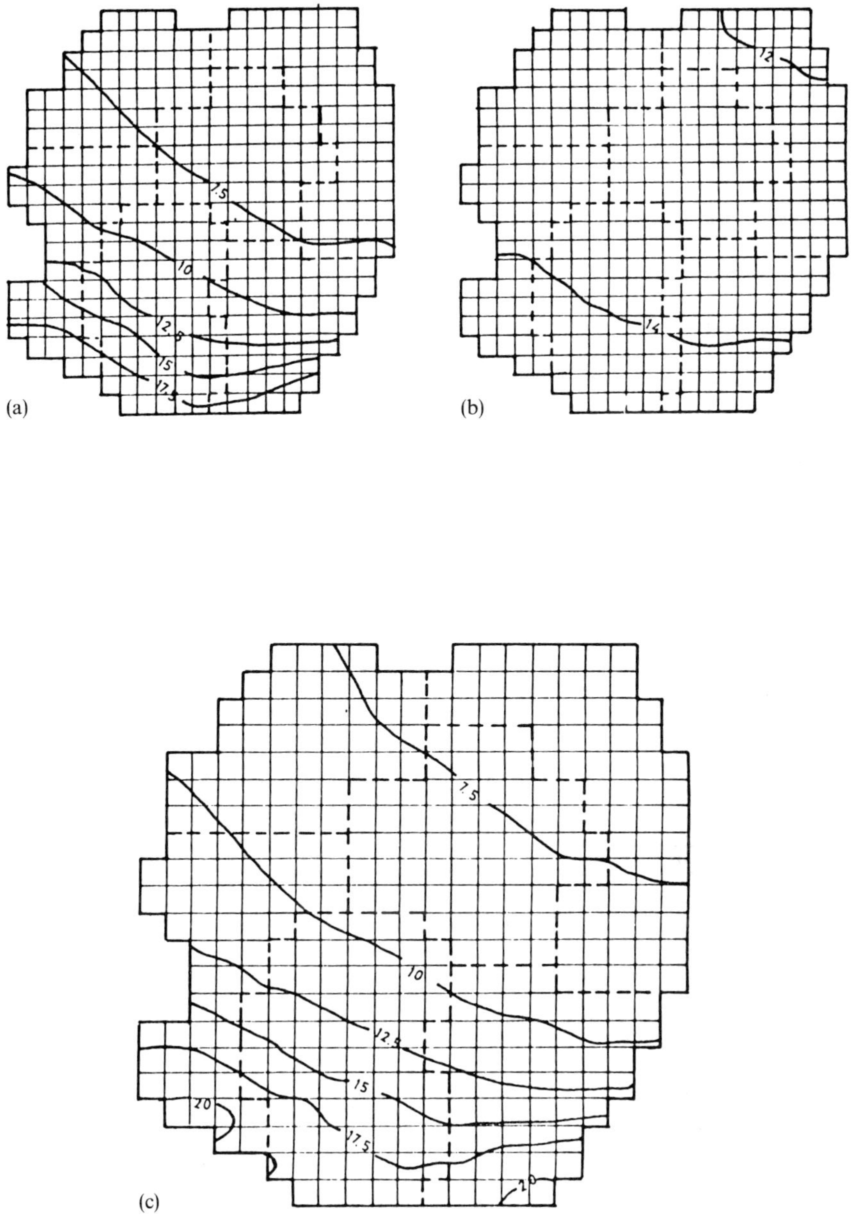

Fig. 7 *Specific loss contours for strengthened masonry buildings for an earthquake with a 200 year return period: (a) damage/usability category II; (b) damage/usability category III; (c) total for damage/usability categories II and III.*

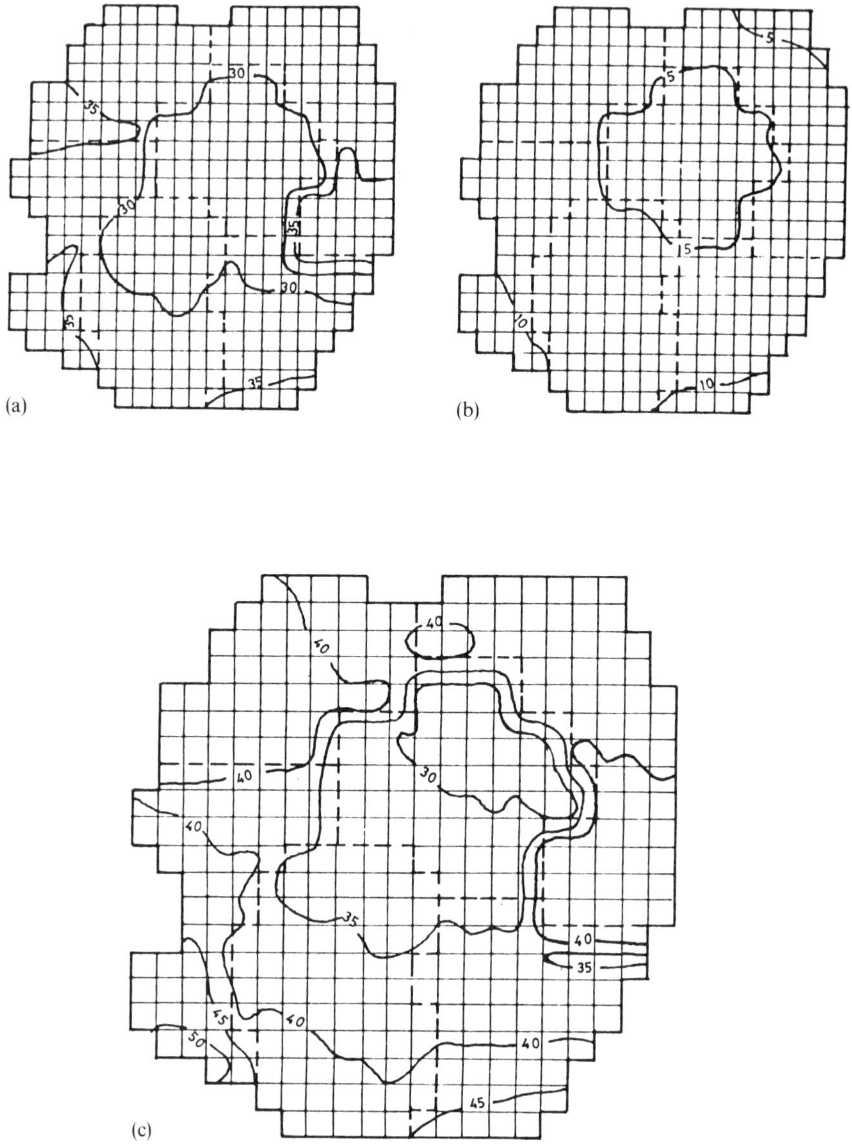

Fig. 8 Cumulative percentage loss for stone masonry, brick masonry, strengthened masonry and reinforced concrete frame buildings for an earthquake with a 200 year return period (scenario A): (a) damage/usability category II; (b) damage/usability category III; (c) total for damage/usability categories II and III.

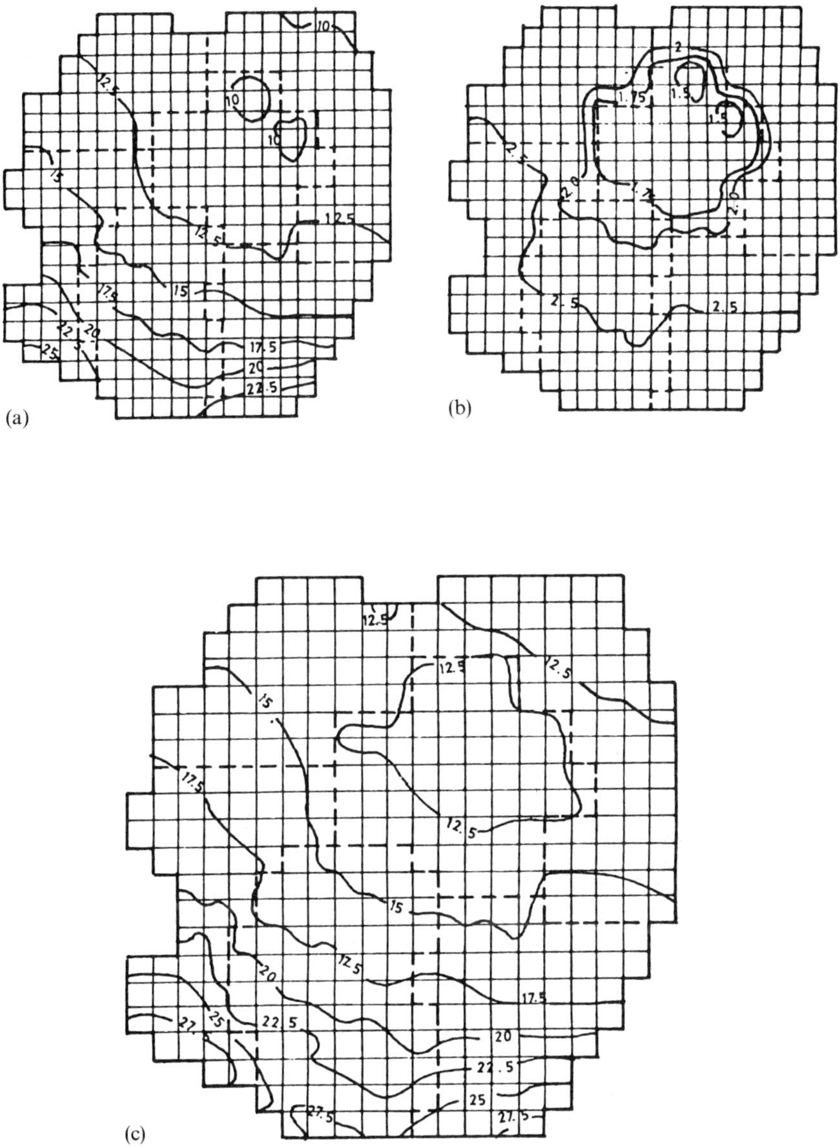

Fig. 9 *Cumulative percentage loss for brick masonry, strengthened masonry and reinforced concrete frame buildings for an earthquake with a 200 year return period (scenario B): (a) damage/usability category II; (b) damage/usability category III; (c) total for damage/usability categories II and III.*

Table 1 *Elements at risk for scenarios A and B*

Commune	*Elements at risk (building class)*								*Total gross area* (m^2)
	Stone masonry		*Brick masonary*		*Strengthened masonry*		*Reinforced concrete*		
	Gross area (m^2)	*% of total*	*Gross area* (m^2)	*% of total*	*Gross area* (m^2)	*% of total*	*Gross area* (m^2)	*% of total*	
Scenario A									
A	208 050	42	59 850	12	122 550	24	108 300	22	498 750
B	472 650	57	103 500	13	203 550	25	41 400	5	821 100
C	224 400	56	45 900	11	79 050	20	53 550	13	402 900
D	315 000	38	78 750	10	151 200	18	286 650	34	831 600
E	280 800	46	81 000	13	180 900	31	62 100	10	604 800
F	390 000	39	101 250	10	210 000	21	303 750	30	1 005 000
Total for region	1 890 900	45	470 250	11	947 250	23	855 750	21	4 164 150
Scenario B									
A	—	—	122 265	25	268 185	54	108 300	21	498 750
B	—	—	245 295	30	534 405	65	41 400	5	821 100
C	—	—	113 220	28	236 130	59	53 550	13	402 900
D	—	—	173 250	21	371 700	45	286 650	34	831 600
E	—	—	165 240	27	377 460	62	62 100	11	604 800
F	—	—	218 250	22	483 000	48	303 750	30	1 005 000
Total for region	—	—	1 037 520	25	2 270 880	55	855 750	20	4 164 150

Table 2 *Comparative total loss for scenarios A and B*

Building class	*Total gross area* (m^2)		*Total loss (%)*					
			Damage/usability category II		*Damage/usability category III*		*Damage/usability category II+III*	
	A	*B*	*A*	*B*	*A*	*B*	*A*	*B*
Stone masonry	1 890 900	—	53.2	—	12.0	—	65.3	—
Brick masonry	470 250	1 037 520	29.5	29.2	6.2	6.2	35.8	35.4
Strengthened masonry	947 250	2 270 850	10.1	9.8	1.3	1.3	11.4	11.1
Reinforced concrete frame	855 750	855 750	9.4	9.4	—	—	9.4	9.4
Total for region	4 164 150		31.7	14.6	6.5	2.3	38.2	16.8

9 are compared in Table 2. It can be seen that there are large differences between the expected cumulative losses expressed as a percentage of the total floor area of the buildings at risk. A fairly small additional investment in preventive measures could have a marked effect on the cumulative loss and significantly reduce the costs of the disaster preparedness programme.

9 Conclusions and recommendations

Seismic activity and the relatively frequent occurrence of large-scale earthquakes in seismically active regions confirm the fact that the population and property in these regions will be increasingly exposed to a large number of small and moderate earthquakes with a significant probability of large catastrophic earthquakes. Therefore rapid development and urbanization, particularly in the earthquake-prone regions of the Mediterranean countries, will result in an increase in vulnerability and seismic risk if appropriate measures to protect human lives and property are not taken. The process of seismic risk reduction and management should start at the level of physical and urban planning and should be consistently implemented.

The vulnerability functions derived for estimating specific loss to elements at risk can be used as basic data for assessing vulnerability and seismic risk in general. The examples presented in this paper provide a simple demonstration of the methodology adopted at the Institute of Earthquake Engineering and Engineering Seismology. An appropriate statistical method must be used to describe the elements at risk and the seismic hazard. Although it is unrealistic to expect elements at risk to be identical in different regions exposed to a seismic environment, the present model can be applied fairly generally. The vulnerability functions derived can be adjusted, if required, and used directly for the same categories of structural types together with a verification of the specific conditions recognized for the regions under consideration.

Since the specific loss is defined as the expected or probable level of loss due to a particular scenario of seismic events, the present model does not estimate the total losses at any given period of time. It assesses the total losses due to a single seismic event expected to take place within a defined return period which is related to the lifetime of the elements at risk. In turn the return period produces an estimated level of hazard at a given site. The probability that this hazard level will not be exceeded is predetermined. There are strong reasons for deriving a model of vulnerability and seismic hazard in this way. Many types of structure are used for housing and public buildings. They are designed in accordance with tradition and with past and recent experience. Sometimes seismic design regulations and codes are obeyed and sometimes they are not. Buildings are constructed with different qualities of workmanship and to different building standards. The above describes the typical characteristics of a region before it is struck by a catastrophic earthquake. A subsequent event will not be faced with the same conditions because normal engineering practice is to repair and strengthen the existing stock of damaged elements at risk in order to achieve economically justifiable safety requirements. It is therefore essential to identify the vulnerability of existing structures. This can be done successfully by performing an immediate damage inventory and systematic statistical classification of data, employing uniform methodological tools to isolate structural types with unacceptable vulnerability levels and replacing them with improved structures which exhibit favourable behaviour and hence a reduced vulnerability level. This can also be done by deriving empirical vulnerability functions for each structural type in accordance with the local site conditions.

Strategies for repairing and strengthening the existing stock of buildings and for planning new development should be based on a comparative evaluation of advance investment in repair and strengthening, thereby increasing the level of seismic safety, versus the probable loss in economic value of property exposed to earthquake hazard. To do this it is necessary to derive theoretical vulnerability functions for strengthened and aseismically constructed structural systems in order to determine the total cost of increasing the level of earthquake protection and to calibrate the ratio of structural to non-structural damage, balancing the associated costs.

The principal advantage of the methodology presented here is that it provides easy access to integrated seismic risk assessment, irrespective of whether or not seismic activity has occurred, by using empirical vulnerability functions for existing traditional buildings and theoretical vulnerability functions for modern aseismic constructions. The other advantage of this methodology is that the data collected and results derived can also be applied to regions of unconfirmed, but probable, seismic activity. The only condition to be met is that the structural types for which the vulnerability functions are derived are similar to those in the region under consideration. If this is not the case, either the empirical or the theoretical vulnerability functions should be adjusted and verified in accordance with the specific conditions of the region under consideration.

The methodology and procedure described here for assessing the damage caused by a major earthquake is accepted for implementation throughout the Balkan countries. It is equally useful in urban and rural areas. The data obtained for earthquake damage and usability of the existing stock of buildings and other structures are of major importance in any planning for reduction of the consequences of earthquakes, post-earthquake rehabilitation and the development of earthquake-stricken regions. Combination of damage data with strong motion data obtained during the earthquake events will significantly improve the store of statistical data needed to derive more reliable and consistent vulnerability functions in the future.

As the amount of statistical data available for earthquakes of different magnitudes and in different regions increases, it may become possible to establish more universal and widely applicable vulnerability relationships which will enable uncertainties in the assessment of seismic loss and the determination of levels of acceptable seismic risk to be significantly reduced.

Bibliography

ALGERMISSEN S. T. (1983): *Seismic Hazard and Risk Assessment: Some Case Studies*, Working Group on Natural Disaster Insurance, Geneva.

FOURNIER D'ALBE E. M. (1982): An approach to earthquake risk management, *Structural Engineer*, **4**, 147–152.

KURIBAYASHI E., UEDA O. and TAZAKI T. (1981): An economic model of long-term effects of earthquake losses, *13th Joint Meeting, US–Japan Panels on Wind and Seismic Effects, Tsukuba, May 1981*, Public Work Research Institute, Ministry of Construction, Japan.

MILUTINOVIC Z. and KAMEDA H. (1983): Equivalent ground acceleration (EGA) as an engineering seismic hazard parameter: development and estimation, *Research Report KUCE 83-ST-01*, School of Civil Engineering, Kyoto University.

PETROVSKI J. (1978): Seismic microzoning and related problems, *The Assessment and Mitigation of Earthquake Risk*, pp. 48–65, UNESCO, Paris.

PETROVSKI J. (1983): Post-earthquake damage evaluation. In *Manual on Damage Evaluation and Assessment of Seismic Resistance of Existing Buildings*, UNDP/UNIDO Project Building Construction under Seismic Conditions in the Balkan Region, Chapter 2, Institute of Earthquake Engineering and Engineering Seismology, Skopje.

PETROVSKI J. *et al.* (1983): Study of expected vulnerability and acceptable seismic risk of six coastal communes and Cetinje for the needs of elaboration of physical plan of Montenegro, Synthesis Report, Vol. I, Seismic hazard, empirical and theoretical vulnerability functions of buildings, and evaluation of observed vulnerability of infrastructure, *Special Report IZIIS 83-97-1*, Institute of Earthquake Engineering and Engineering Seismology, Skopje (in Serbo-Croatian).

PETROVSKI J., STANKOVIC V., NOCEVSKI N., RISTIC D. and PAVICEVIC B. S. (1984): Study of expected vulnerability and acceptable seismic risk of six coastal communes and Cetinje, Synthesis Report, Vol. II, Analysis of expected vulnerability and determination of the level of acceptable seismic risk, *Special Report IZIIS 83-97-2*, Institute of Earthquake Engineering and Engineering Seismology, Skopje (in Serbo-Croatian).

PETROVSKI J. *et al.* (1984): Development of empirical and theoretical vulnerability and seismic risk models, *Proceedings of the 8th World Conference on Earthquake Engineering*.

UNDRO (1979): Natural Disaster and Vulnerability Analysis, *Report of Expert Group Meeting*, UNDRO.

J. H. WIGGINS Co. (1980): Assessment of damageability for existing buildings in a natural hazard environment, Vols I and II, Methodology, *Technical Report 80-1332-1*, J. H. Wiggins Co., Redondo Beach, CA.

IRAS: An Expert System for Earthquake Insurance and Investment Risk Assessment

H. C. SHAH, F. S. WONG, W. DONG, M. LAMARRE AND G. H. MIYASATO

Department of Civil Engineering, Stanford University, Stanford, CA, USA

Abstract The development of the Insurance and Investment Risk Analysis System (IRAS) which provides consultation for the earthquake insurance and investment banking industries is summarized. The features of this system are described including interactive input–output facilities, graphic data retrieval, hierarchical knowledge-based management, an integrated system of independent program modules, combinations of backward-chaining and forward-chaining inference mechanisms, and approximate reasoning schemes based on fuzzy set theory to deal with linguistic and/or incomplete information.

1 Introduction

Recent advances in earthquake engineering have permitted experts to estimate earthquake hazard, vulnerability and risk rationally, but this expertise is not readily available to managers in insurance and banking industries because there are problems in defining and obtaining the relevant data for seismic risk evaluation and applying these data for decision making. Thus the purpose of the present project was to develop a knowledge-based expert system, known as the Insurance and Investment Risk Analysis System (IRAS), on the seismic vulnerability of existing buildings for use by real-estate investment analysts, insurance underwriters and reinsurers, decision makers and portfolio managers, and structural engineers and appraisers in evaluating and revising investment strategies and insurance policies or premiums.

There are three specific expert systems in IRAS. The first system evaluates seismic hazard for a given site. It integrates an existing database of past geological and seismological information, which includes the location of seismic sources, the recurrence relationships for all these sources and any secondary hazard information that is available in the literature. The second system evaluates the seismic vulnerability of a group or class of buildings in a given region whose seismic hazard has been evaluated by the first expert system. This system can also be used for rapid identification of high-risk buildings in a region. It relies only on readily available information about the buildings under investigation; no detailed engineering drawing or other input is needed. The third system evaluates detailed vulnerability and risk for a specific building which either has been identified as a high-risk building by the previous expert system or is of particular interest to the analyst.

IRAS has several features. First, it adopts the commercial software I/O PRO as the main input–output facilitator. The screen development system facilitates the creation of text and graphic screens for use as the input and output media for interactive programs. The slides displayed on the screen are used to communicate with the user for input and output and explanations. Input data formats can be numerical, linguistic or graphical depending on the context. Second, the control strategy for the inference mechanism in IRAS is a combination of backward chaining (goal driven) and forward chaining (data driven). The system uses backward chaining to satisfy diverse goals (enquiries). However, when the goal is specified the system uses forward chaining to collect the relevant data. Since goal specification significantly reduces the search space, only minimum search effort is required. Third, unlike conventional rule-based systems, IRAS recognizes the fact that seismic risk evaluation needs both judgmental expertise and well-established mathematical procedures. Hence IRAS incorporates both rule-based systems and algorithmic programs which saves a great deal of computation. Fourth, to increase the ease of upgrading IRAS and to ease the restriction of internal memory, the submodules of IRAS are written as independent programs. The programs are compiled independently and are then called into memory when needed by the driver. Fifth, IRAS adopts the current probabilistic approach for hazard analysis to handle uncertainties in the prediction of ground shaking for the site. The design detail and construction quality of a building will significantly influence its performance during earthquakes and must be identified in order to obtain a reasonable evaluation of the vulnerability of the building. To reflect the judgmental knowledge of the effect of different factors on building damage, IRAS uses an uncertainty model based on fuzzy set theory.

IRAS is designed in such a way that the knowledge bases can be updated as new information is made available. Furthermore, the systems are user friendly, and hence their repeated use can make them a standard for evaluating seismic hazard and risk for the banking and insurance industries.

2 Domain knowledge

Seismic risk is defined as the likelihood of loss due to earthquakes and involves four basic components: hazards, exposure, vulnerability and location. These factors are defined further below (Miyasato *et al.*, 1986).

(1) The **hazards** or dangerous situations can be classified as follows:
   (i) primary hazards (fault break, ground vibration);
   (ii) secondary hazards which are potentially dangerous situations triggered by the primary hazards (for example, a fault break can cause a tsunami, or ground shaking can result in foundation settlement, foundation failure, liquefaction, landslides etc.);
   (iii) tertiary hazards produced by flooding by dam break, fire following an earthquake and the like.

   All these hazards lead to damage and loss. They can be expressed in terms of severity, frequency and location.
(2) The **exposure** is defined as the value of the structures and contents, interruption of business, loss of life etc.
(3) The **vulnerability** is defined as the sensitivity of the exposure to the hazard(s) and the location relative to the hazard(s).
(4) The **location** is defined as the position of the exposure relative to the hazard.

Losses resulting from seismic hazard are numerous and can be categorized as follows: life and injury; property; interruption of business; lost opportunities; contents; tax base; other losses. A seismic risk analysis requires the identification of losses to be studied as well as the identification of hazard exposures and their locations and vulnerability.

For the purpose of the insurance and the real-estate industries, property losses are the major concern. Property loss is usually measured by the damage ratio which is defined as the repair cost of the damaged facility divided by the initial cost of the facility. Owing to uncertainties in predicting structural behaviour during future earthquakes, the current practice of the insurance industry in California is to use probable maximum loss (PML) (Steinbrugge, 1982) as the basis for premium calculation. PML is defined as the damage ratio such that during the "maximum probable" earthquake nine out of ten buildings will experience damage less than the value given by the PML.

The PML does not take into account the randomness of earthquake occurrence with respect to time, location or earthquake size. In order to reflect the uncertain nature of earthquake occurrence a second index, called the damage threshold (DT), is used which combines the uncertain response of the building with the random occurrence of future earthquakes (Chiang *et al.*, 1984; Shah and Dong, 1984; Applied Technology Council, 1986). Both indices are used in IRAS.

IRAS is divided into three subsystems corresponding to the major components of seismic risk: SHES, SRES1 and SRES2. The seismic hazard evaluation system (SHES) combines hazard and location components to obtain the seismic hazard estimation. The main flow chart for SHES is shown in Fig. 1. The seismic risk

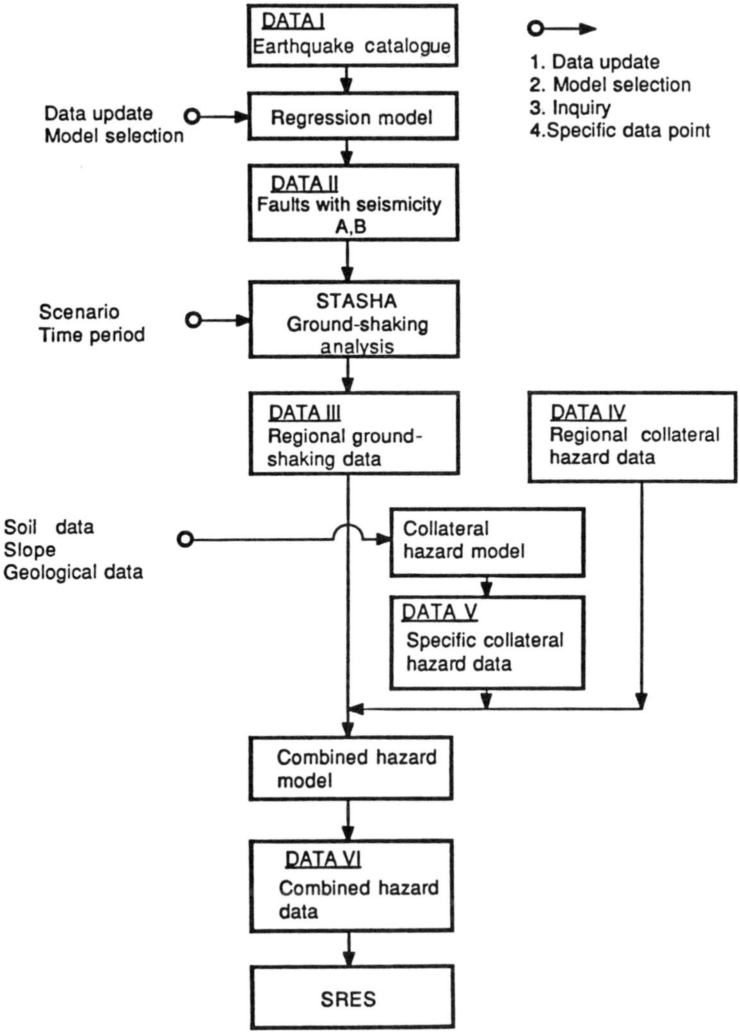

Fig. 1 *Seismic hazard evaluation system (SHES).*

evaluation system (SRES1) is used to screen the property loss from exposure and vulnerability of the building. Only building type (classification) is required at this level. The flow chart for SRES1 is shown in Fig. 2. Figure 3 shows the flow chart of SRES2, which performs the second level of seismic risk evaluation taking into account specific information about the buildings. Data management and the inference mechanism of these subsystems will be described in the following sections.

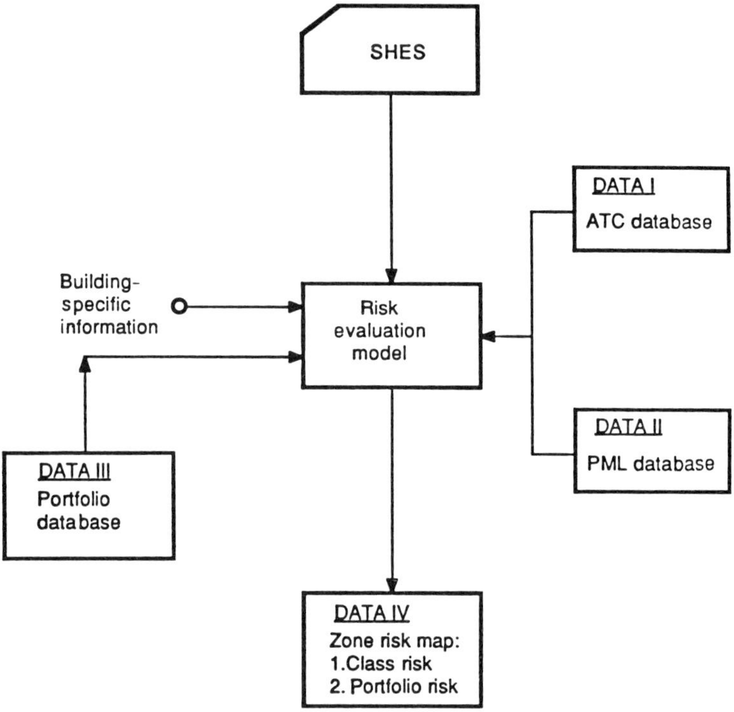

Fig. 2 *Seismic risk evaluation system (SRES1).*

3 Inference mechanism (inference engine)

An inference engine incorporates reasoning methods which act upon input data and knowledge from the knowledge base to solve the desired problem and produce an explanation when requested. Control strategy for the inference engine can be forward chaining, backward chaining or a mixture of both.

In the IRAS application the system should be able to satisfy diverse goals (enquiries) such as "what is the real-estate investment portfolio risk for a given region due to a catastrophic earthquake?" or "what is the PML of a particular building due to all contributing fault seismicity?" The goal specifies the reasoning path that should be pursued. Hence it is natural that backward chaining (goal driven) should be used. However, when the goal is specified and the reasoning path to achieve this goal is identified, the system will use forward chaining (data driven) to collect the relevant data either by querying the user or by searching and retrieving them from the knowledge base. Thus the control mechanism is a combination of backward chaining and forward chaining. Since goal specification significantly reduces the search space, only a minimum search effort is required.

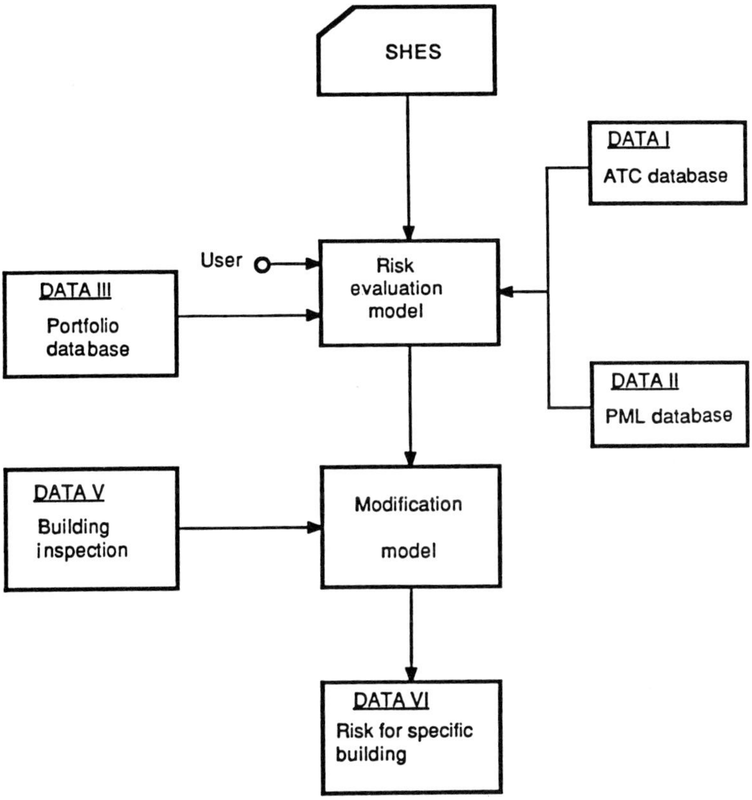

Fig. 3 *Seismic risk evaluation system (SRES2).*

4 Knowledge base (database)

The knowledge base (database) for IRAS consists of raw data, production rules, engineering and analysis programs and approximate reasoning schemes. Unlike conventional rule-based systems which use if–then rules only, IRAS recognizes the fact that seismic risk evaluation needs both judgmental expertise and well-established mathematical procedures. Hence, IRAS incorporates both if–then production rules and algorithmic programs. For instance, model selection depends heavily on the expert's subjective judgment, and if–then rules are suitable for guiding the user to select the appropriate model. After the model is selected, the relevant procedures are executed using algorithmic programs.

The earthquake database is provided by the National Geophysical Data Center. The database contains about 27 000 earthquakes above magnitude 3.0 recorded since 1900. A method of converting the different magnitude and intensity scales encountered in the database into a uniform scale, the local magnitude scale, is being developed. Significant earthquakes that occurred before 1900 will also be included in the database to improve the modelling of occurrence of larger earthquakes. The database is complete only for earthquakes above magnitude 4.0 after 1933 and in several instances the occurrence of earthquakes in swarms makes it difficult to differentiate slightly differing recordings of the same earthquake. These factors will be considered in our modelling of earthquake recurrences. Each earthquake is associated with the source having the closest segment to its epicentre. This information is stored in the database and will be used to determine all the earthquakes associated with a source. The database can be updated any time to account for new earthquakes. The system is designed such that the USGS and the National Geophysical Data Center formats can easily be read to include new earthquakes in the existing database.

The source database was constructed from the California Division of Mines and Geology fault map drawn at a scale of 1:750 000. Their map was digitized with a GTCO digitizing tablet with a resolution of 0.1 mm. Several intersections of degrees of longitude and latitude as well as points along the faults were input to a program that transformed the cartesian coordinates into degrees of latitude and longitude. The faults are composed of several segments whose end-points are stored in the database. Once determined, the recurrence and attenuation models and their parameters will also be stored in the database. To determine the recurrence model on a source all the earthquakes that belong to that source are used. The recurrence models to be considered are discussed below.

4.1 Poisson model

The Poisson recurrence model assumes a memoryless process and is characterized by the following equation:

$$p(m_c, T) = 1 - \exp\{\gamma_t T G_M(m_c)\}$$

where $p(m_c, T)$ is the probability that a critical magnitude level m_c will be exceeded at least once in the next T years, γ_t is the expected number of events per year that have a magnitude greater than or equal to m_t, m_t is the threshold magnitude of significant events which release substantial stress along the fault, effectively restarting the strain accumulation process, and $G_M(m)$ is the complementary cumulative distribution of all magnitudes greater than m_t ($G_M(m_t) = 1.0$).

The magnitude distribution is very often assumed to be the Gutenberg–Richter distribution which is represented by the following equation in the untruncated case:

$$G_M(m) = \exp\{-\beta(m - m_t)\} \qquad m \geqslant m_t$$

However, the Poisson model allows any magnitude distribution to be used instead of the Gutenberg–Richter relation. On some faults the distribution of magnitudes may peak near a large characteristic-magnitude level and the characteristic-size model should be used. This model is obtained by generalizing the Gutenberg–Richter relationship using the Weibull distribution as follows:

$$G_M(m) = \exp[-\{\beta(m-m_t)\}^{1/V_M}] \qquad m \geqslant m_t$$

where

$$V_M = \frac{\{\mathrm{var}(M)\}^{1/2}}{E(M)-m_t}$$

and

$$\beta \approx \frac{1}{E(M)-m_t}$$

4.2 Characteristic-time model

The Poisson model implies an interevent time given by the equation

$$G_\tau(t) = \exp(-\lambda t) \qquad t \geqslant 0$$

where $G_\tau(t)$ is the complementary cumulative distribution function of the interevent times and λ is the mean annual rate of events of magnitude greater than m_t. If the interevent times on a fault show less variability than the memoryless model implies, the characteristic-time model can be used instead to describe the distribution of the interevent times. This model is obtained by generalizing the previous equation using the Weibull function as follows:

$$G_\tau(t) = \exp\{-(\lambda t)^{1/V_\tau}\}$$

where

$$V_\tau = \frac{\{\mathrm{var}(\tau)\}^{1/2}}{E(\tau)}$$

and

$$\lambda \approx \frac{1}{E(\tau)}$$

With the Poisson interevent times the hazard function, which is the probability that an event will occur in the next infinitesimal time interval given that it has not happened in the last t_0 years, is a constant λ. The Weibull distribution has an increasing hazard function with t_0, which is more realistic in some cases. With the characteristic-time model the probability of exceeding a magnitude level increases with the elapsed time t_0 since the last significant event.

4.3 Time-predictable model

If there is a positive correlation between the interevent time and the preceding magnitude level m_0, a time-predictable model can be assigned to the source. To characterize the time-predictable model, both the conditional distribution $G_{\tau|M_0}(t|M_0)$ of the interevent times given the value of the preceding magnitude and the marginal distribution $G_M(m)$ of the independent successive magnitudes must be specified. For instance, given m_0, the interevent times can be assigned a Weibull distribution with parameters $E(\tau|M_0)$ and $V_{\tau|M_0}$ and the magnitudes can be assigned a Weibull distribution with parameters $E(M)$ and V_M.

4.4 Slip-predictable model

If there is a stochastic dependence between the interevent time τ and the subsequent magnitude m_1, a slip-predictable model can be assigned to the source. The slip-predictable model is specified by assigning both the conditional distribution $G_{M_1|\tau}(m_1|t)$ of the subsequent magnitude given the interevent time and the marginal distribution $G_\tau(t)$ of the independent successive interevent times. The Poisson model is adequate for most cases (Cornell, 1986).

4.5 Testing the recurrence models

Two methods are available for testing the Poisson model: the plot method and the goodness-of-fit method. If the plot of the Poisson quartiles versus the sorted interevent times is a straight line, the assumption of a Poisson model is valid. If the points do not lie on a straight line, the Poisson hypothesis is rejected. Since it is difficult to decide whether or not the plot is a straight line, the plot method should be used carefully unless the line is obviously straight or random.

To verify the Poisson assumption of earthquake occurrences or the Gutenberg–Richter assumption of exponentially distributed magnitudes, the goodness-of-fit test can be used as follows. Given that the expected count in a certain category i equals $\tilde{n}_i$, the probability that the count is less than or equal to the observed count n_i is

$$P(N_i \leqslant n_i) = \sum_{k=0}^{n_i} \frac{\tilde{n}_i^{\,k} \exp(-\tilde{n}_i)}{k!} = \alpha_i$$

To verify the Poisson assumption of earthquake occurrences on a fault, the expected and observed counts of earthquakes on this fault in different time periods are compared. The exponential distribution of magnitudes on a fault is verified by comparing the expected and observed counts on this fault for different magnitude intervals. Very low and very high values of α_i indicate that the expected count is too high or too low respectively and the assumption tested is rejected.

To determine whether the characteristic-size model or the Poisson model is the best for a source, a test is performed on the statistic

$$V_M = \frac{\{\mathrm{var}(M)\}^{1/2}}{E(M) - m_t}$$

If it is significantly different from 1.0, the Gutenberg–Richter assumption is rejected and the characteristic-size model using a Weibull distribution is recommended. Similarly, to determine whether the characteristic-time model or the Poisson model is the best for a source, a test is performed on the statistic

$$V_\tau = \frac{\{\mathrm{var}(\tau)\}^{1/2}}{E(\tau)}$$

If it is significantly different from 1.0, the Poisson assumption is rejected and the characteristic-time model using a Weibull distribution is recommended.

The slip- or time-predictable assumptions can be verified using the following tests. If the correlation between the interevent time and the preceding magnitude level is greater than zero ($\rho_{\tau,M_0} > 0$), the time-predictable model is recommended. If there is a positive correlation between the interevent times τ and the subsequent magnitudes M_1 ($\rho_{M_1,\tau} > 0$), the slip-predictable model is recommended. If both correlations are greater than zero then a combined slip- and time-predictable model is recommended.

The recurrence model for a source can also be determined by using rules recreating the opinions of experts. These rules facilitate the task of determining which recurrence model is most appropriate for a source because they bypass the necessity for performing all the statistical tests discussed above. A few examples of such rules are given below.

RULE 1:

IF the mean interevent time between significant events is greater than the elapsed time since the last such event
THEN the Poisson model might be adequate.

RULE 2:

IF the Poisson model might be adequate
AND the mean interevent time between significant events is greater than the length of the historical record
THEN the Poisson model is adequate.

RULE 3:

IF the hazard is controlled by a single fault
AND the time since the last significant event is greater than the average time between such events on that fault
AND a strong characteristic-time behaviour applies to that fault
THEN the Poisson model is not adequate.

5 Integrating independent programs

A common practice in programming is to have a main driver and many subroutines.

The driver and subroutines are compiled into a global executable program. However, when the problem to be solved is complex, many subroutines and submodules are needed. The size of the program increases rapidly and soon the capacity of the internal memory of a microcomputer is exceeded. It is then necessary to rely on complex input–output manipulation and peripheral storage to fit the program into the computer. Furthermore, when any submodule of the program needs to be changed because of technological or engineering advances, the relevant routines must be changed and recompiled. The fitting must be reconstructed.

To increase the ease of upgrading IRAS and to ease the restriction of internal memory on the IBM AT, the submodules of the systems are written as independent programs. Most of these programs have been developed during the past 10 years by staff and students at the Blume Earthquake Engineering Center, and they are simply ported to the microcomputer. The programs are compiled independently and individually and are then called into memory when needed by the driver, in a similar way to a subroutine. The retrieval, execution and return of the external programs is easily achieved on the IBM AT by using the interrupt feature of DOS. Each external program can be as large as the total internal memory of the AT (currently 640 k).

6 Uncertainty treatment

As mentioned in previous sections, uncertainties are involved at each stage of the evaluation process. Earthquake occurrence and size are random in nature. For this type of uncertainty, the probabilistic approach has been well established and the data in California are good enough to support estimation using this approach. Hence IRAS adopts the current probabilistic approach for hazard analysis to handle uncertainties in the prediction of ground shaking for the site. The program STASHA, developed at Stanford University for hazard analysis, was incorporated into IRAS using the approach described in the previous section.

There is yet another type of uncertainty in the evaluation which cannot be handled using probabilistic methods. Design detail and construction quality will affect the performance of a building and need to be taken into account in evaluation of its vulnerability. The damage degree will vary over a wide range from bad engineering design to good engineering design. All these factors will significantly influence the building performance during earthquakes and must be identified in order to obtain a reasonable evaluation. When the user fails to answer the enquiry on these factors from the systems, it is then expected that the system will give an answer with a wider spread owing to the larger uncertainty. Because data regarding damage from diverse building types are scarce and are not sufficient to support a probability distribution, IRAS uses an uncertainty model based on fuzzy set theory to reflect the judgmental knowledge of the effect of different factors on building damage.

Fuzzy sets with different membership functions are used to represent the prior information on these effects. Some examples are shown in Fig. 4. Whenever the

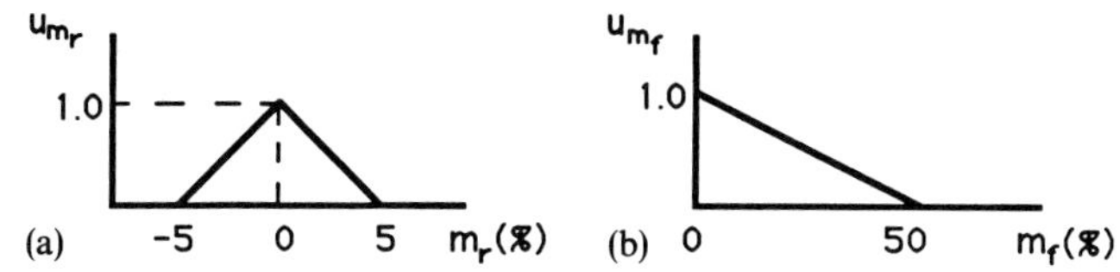

Fig. 4 *Fuzzy representations of modification factors for (a) a roof system and (b) fault rupture.*

response to a query for data is unknown (the default), the system will use the fuzzy set instead of a crisp number to count its effect. The vertex method (Dong and Shah, 1987; Dong and Wong, 1987) is used to combine all these effects and to calculate the total effect, resulting in a certainty factor which reflects the degree of uncertainty. When the system gives the evaluation result, it also indicates the reliability of the result (certainty factor) and how the reliability can be improved.

7 *Input–output facilities*

IRAS adopts the commercial software I/O PRO, developed by MEF Environmental (MEF, 1985), as the main input–output facility. I/O PRO is a modular set of software development tools and utilities which together create a high productivity environment for FORTRAN, C and Pascal programmers. The screen development system facilitates creation of text and graphic screens for use as the input and output media for interactive programs. The slides displayed on the screen are used to communicate with the user for input and output and explanations if requested. Input data formats can be numerical, linguistic or graphical, depending on the context.

In addition to the interactive mode, the user can also choose the batch mode in which all data are read in together using a format such as the Lotus 1-2-3 spreadsheet or Dbase III. This mode facilitates the data transfer from insurance and investment banking company databases.

In order to display the regional risk, IRAS also incorporates another commercial software, ATLAS (Strategic Locations Planning, 1985), to show the thematic map of regional risk. All input–output options are built into the master program and can be exercised according to the user's goal and decision needs.

8 *Conclusion*

The development of the seismic risk analysis system IRAS for the insurance and real-estate investment industries is summarized in this paper. The main framework of the system has been completed and demonstrated to clients and users. The response from potential users has been excellent. Next, detailed data will be filled in and decision models will be incorporated into the system. The system will be operational by the end of 1987 after overall validation.

IRAS has been developed for buildings located in California, which is the area of its sponsors' main interest. However, in view of the modularity and flexibility of its design, IRAS can be readily adapted to other regions of the world when the appropriate data/knowledge bases are available.

Acknowledgments

The authors wish to express their sincere gratitude to Aetna and Salomon Brothers for their support, and to Peter Wachtel, David Gold, Bill Hettinger, Kevin Schmitt, Lance Manual and Jong-Eup Kim for their cooperation.

References

APPLIED TECHNOLOGY COUNCIL (1986): *Earthquake Damage Evaluation Data (ATC-13)*, Applied Technology Council, Redwood City, CA.

CHIANG W. L., GUIDI G. A., MORTGAT C. P., SCHOOF C. C. and SHAH H. C. (1984): Computer programs for seismic hazard analysis, *Report 62*, John A. Blume Earthquake Engineering Center, Stanford University, Stanford, CA.

CORNELL C. A. (1986): *Applicability of the Poisson Earthquake Occurrence Model, EPRI NP-4770, Research Project P101-38*, EPRI, Palo Alto, CA.

DONG W. and SHAH H. (1987): Vertex method for computing the function of fuzzy variables, *Fuzzy Sets and Systems*, **24** (1), 65–78.

DONG W. and WONG S. (1987): Fuzzy weighted averages and implementation of the extension principle, *Fuzzy Sets and Systems*, **21** (2), 183–199.

MEF ENVIRONMENTAL INC. (1985): *I/O PRO—Screen Development System and Utilities, Version 1.1*, MEF Environmental Inc., Austin, TX.

MIYASATO G., DONG W., LEVITT R. and BOISSONNADE A. (1986): Implementation of a knowledge based seismic risk evaluation system on microcomputers, *Artificial Intelligence in Engineering*, **1** (1), 29–35.

SHAH H. C. and DONG W. (1984): A reevaluation of the current seismic hazard assessment methodologies, *Proceedings of the 8th World Conference of Earthquake Engineering*, Vol. 7, pp. 247–254, Prentice-Hall, Englewood Cliffs, NJ.

STEINBRUGGE K. V. (1982): *Earthquakes, Volcanoes, and Tsunamis—An Anatomy of Hazards*, Skandia America Group, New York.

STRATEGIC LOCATIONS PLANNING (1985): *ATLAS-AMP-EGA*, Strategic Locations Planning, San Jose, CA.

The Assessment of Seismic Risk

E. M. FOURNIER D'ALBE†

4 rue de la Boutinaye, Dormont, 27920 St Pierre de Bailleul, France

Abstract The systematic assessment and recording of damage to large numbers of buildings of various structural types after recent earthquakes in Bucharest, Romania (March 1977), and in the Republic of Montenegro, Yugoslavia (April 1979), makes it possible to derive empirical vulnerability functions for these types of structure. In this paper the results of an analysis relating observed degree of damage to (a) macroseismic intensity, (b) estimated peak ground acceleration and (c) epicentral distance are given. The limitations of these approaches to the estimation of specific risk are discussed.

1 Introduction

Estimates of seismic risk (the probability of loss through earthquake action in a given area in a given period of time) can be of considerable use to physical and economic planners concerned with investment and settlement in seismic areas, to those concerned with insurance against such losses and to civil defence officials whose task it is to prepare plans for rescue, relief and rehabilitation after a possible future earthquake disaster. At government level, such estimates may help to determine policy with regard to the human and financial resources that should be invested in scientific and technical research, the control of building design, inspection of construction etc.

Although there have been many studies of the assessment of seismic hazard (the probability of occurrence of seismic ground motion at any site), there have been relatively few attempts to assess the probability of damage to or destruction of existing buildings, or the attendant human, social and economic losses. This is probably due to the lack of precise information on the vulnerability of existing

† deceased.

buildings, i.e. on the degree of damage or proportional loss of value that they are likely to suffer when subjected to ground motion of various intensities. Unless the design of a building has been the subject of detailed dynamic analysis and the materials used in its construction have been carefully tested, it is very difficult to estimate its vulnerability. Therefore the only clues to the probable behaviour of the great majority of existing buildings in future earthquakes are to be found in the analysis of the behaviour of similar buildings in past earthquakes.

The systematic observation and analysis of earthquake damage for this purpose is a recent development. In this paper we shall use the damage data which exist for two recent earthquakes (those of 4 March 1977 in Romania and of 15 April 1979 in Montenegro, Yugoslavia) to illustrate a method of seismic risk assessment based on observed vulnerability.

Before proceeding further it may be useful to recall the UNDRO definitions (UNDRO, 1979) of the terms vulnerability, hazard and risk on which the present work is based. The **vulnerability** of a structure, or of a set of structures of the same given type, is the degree of damage or loss which results from the occurrence of seismic ground motion of given intensity. The **vulnerability function** of a structure is that function which describes, graphically or mathematically, its vulnerability to various intensities of ground motion, where the intensities are preferably expressed by values of some physical parameter such as acceleration but in some cases by a degree on a macroseismic scale. The **seismic hazard** at any site is expressed by the probability distribution of the intensities of seismic ground motion at the site in a given period of time. It should therefore be distinguished from the **seismicity** of an area, which is expressed by the frequency and magnitude of earthquakes in that area. In the event of an earthquake, the degree of damage or loss at a given site will be determined by the intensity of ground motion at the site and by the vulnerability of the buildings (if any) at the site to motion of that intensity. In the intervals between earthquakes, the specific risk to these buildings will be determined by their vulnerability functions and by the seismic hazard at the site. By **specific risk** we mean the expected loss in a given lapse of time, expressed as a proportion of the value (or replacement cost) of the buildings.

2 *Basic methodology*

No two buildings are identical. In order to be able to derive information of prognostic value from the inspection of earthquake damage, buildings must therefore be classified into a limited number of structural types. The definition of each type must be narrow enough to ensure some degree of uniformity in the seismic response of all buildings within the type but wide enough to ensure that the number of buildings in each type is large enough for statistical analysis.

No two buildings are damaged in exactly the same way. It is therefore necessary to define a set of criteria by which the observed damage can be categorized, so that the degree of damage can be related to the degree of loss (e.g. the cost of repair expressed as a proportion of value or of replacement cost).

Lastly, some way has to be found of expressing, as accurately as possible, the intensity of ground motion at the site of the building. As we shall see, this is perhaps the most difficult problem of all.

2.1 Classification of buildings

After the earthquake of 4 March 1977 in Romania, a detailed survey of the damage suffered by some 18 000 buildings in the city of Bucharest was undertaken under the auspices of the Romanian National Council for Science and Technology (Sandi and Vasilescu, 1982). The buildings were divided into five categories according to the construction system:

- C1 low quality construction (adobe like);
- C2 masonry-bearing walls with flexible floors;
- C3 masonry-bearing walls with rigid (reinforced concrete) floors;
- C4 reinforced concrete frame structures;
- C5 reinforced concrete bearing-wall structures.

Categories C2 and C3, which represented the bulk of the buildings surveyed, were subdivided according to age (pre-1940 and post-1940).

After the earthquake of 15 April 1979, an extensive survey was made of damage to over 40 000 buildings in Montenegro, and the data were collected and compiled by the Institute of Earthquake Engineering and Engineering Seismology, University "Kiril i Metody", Skopje. The 39 000 inspected buildings of less than four storeys were classified into five categories:

(1) stone masonry;
(2) brick masonry;
(3) strengthened masonry (timber belts, steel ties or reinforced concrete belts);
(4) reinforced concrete frame structures;
(5) reinforced concrete shear-wall structures.

These five categories were broken down into numerous subcategories for detailed engineering studies, which do not concern us here.

2.2 Quantification of damage

Because earthquake damage to buildings is often extremely complex and because post-earthquake damage surveys usually have to be carried out by large numbers of inspection teams working independently of each other, it is important to define standard methods of assessing the degree of damage to each building.

In the damage survey carried out in Romania after the earthquake of March

1977, standard scales of degree of damage were adopted for each type of construction (Sandi and Vasilescu, 1982). As an example, the damage scale for reinforced concrete frame structures is reproduced in Table 1. The degree of damage of the whole building was taken to be whichever was the greatest of the figures given in Table 1. In the case of the earthquake of April 1979 in Montenegro the scale

Table 1 *Degrees of damage for reinforced concrete frame structures (Romania 1977)*

Columns		*Beams*		*Infill masonry*	
Not affected	0	Not affected	0	Not affected	0
Cracked	2	Slightly affected	1	Boundary cracks	0
Severe cracking	4	Cracked	2	Cracking	1.5
Crushed concrete	4	Severe cracking	3	Severe cracking	2
Crushed reinforcement	4	Failure	4	Dislocation	2.5
				Collapse	3

Table 2 *Degrees of damage of buildings in Montenegro, 1979*

Degree of damage	*Phenomena observed*
1 (none)	No visible damage to the structural elements; possible fine cracks in wall and ceiling mortar; barely visible non-structural and structural damage
2 (slight)	Cracks in wall and ceiling mortar; falling of large patches of mortar from wall and ceiling surface; considerable cracks in or partial failure of chimneys, attics and gable walls; disturbance, partial sliding, sliding and collapse of roof coverings; cracks in structural members
3 (moderate)	Diagonal or other cracks in structural walls, walls between windows and similar structural elements; large cracks in reinforced concrete structural members (columns, beams, reinforced concrete walls); partially failed or failed chimneys, attics or gable walls; disturbance, sliding and collapse of roof covering
4 (heavy)	Large cracks with or without detachment of walls, with crushing of materials; large cracks with crushed wall material between windows and similar elements of structural walls; large cracks with slight dislocation of reinforced concrete structural elements (columns, beams and reinforced concrete walls); slight dislocation of structural elements and the whole building
5 (severe)	Structural members and their connections undergo extreme damage and dislocation; many crushed structural elements; substantial dislocation of the entire building and damage to roof structure; partial or complete failure

shown in Table 2 was adopted for all types of buildings (Petrovski *et al.*, 1984). It should be noted that degree 1 on the Montenegro scale corresponds to degree 0 on the Romanian scale; degree 5 is the same on both scales.

2.3 Quantification of specific loss

Except in the case of severe structural damage or collapse, when the specific loss can be taken as 100%, it is not easy to define the relationship between degree of damage and specific loss (cost of repair expressed as a proportion of the estimated value or cost of replacement) since this will depend on many local factors that are likely to vary widely from country to country and from case to case.

The simplest assumption would be that the specific loss is linearly proportional to the degree of damage, so that in the case of Romania, for instance, degree 1 would correspond to a specific loss of 20%, degree 2 to a specific loss of 40% and so on. However, these would almost certainly be overestimates. For the purpose of demonstrating a method of risk assessment, we shall assume here that the specific loss is proportional to the square of the degree of damage. This gives us the correspondence shown in Table 3. It must be emphasized that the relationships shown in Table 3 are purely arbitrary and are used only to illustrate the method. Actual losses and repair costs may differ significantly from them.

Table 3 *Relation between degree of damage and specific loss*

Degree of damage	*Specific loss (%)*	
	Romanian case	*Yugoslav case*
1	4	0
2	16	6
3	36	25
4	64	56
5	100	100

2.4 Assessment of ground motion intensity

For the purposes of risk assessment, our main objective in analysing observed damage is to derive empirical relationships (vulnerability functions) between the degree of damage and some parameter representing the "intensity" of seismic ground motion.

Two accelerograph records were obtained in Bucharest during the March 1977 earthquake, one in the basement of a single-storey building and the other on the upper floor of a 10-storey building about 2 km away (Sandi and Vasilescu, 1982). It is clearly not possible to estimate the peak ground acceleration in all parts of the city with different local soil conditions from only two records. The urban area was therefore divided into squares of dimensions 1 km × 1 km, and the Medvedev–Sponheuer–Karnick (MSK) intensity in each square was assessed from the observed damage to all types of buildings. It was of course realized that this method of intensity assessment suffers from the handicap that the two variables, degree of damage and intensity, cannot be considered as independent. However, it is possible to use this method to compare the vulnerabilities of different types of buildings, and this was in fact done by Sandi and Vasilescu (1982).

During the Montenegro earthquake of 15 April 1979, 25 accelerograph records were obtained of the magnitude 7.2 event, of which six showed peak accelerations exceeding 0.05g, the lower limit at which significant damage occurred. On the basis of these recordings, and of others obtained during aftershocks, maps were prepared showing the estimated values of peak ground acceleration throughout the macroseismic area. In the analysis presented here these values have been taken as measures of the intensity of ground motion. In view of the fact that these estimated values do not take account of the possible influence of local soil conditions, the buildings in each construction type have been subdivided according to whether they were located on rock or on alluvial ground.

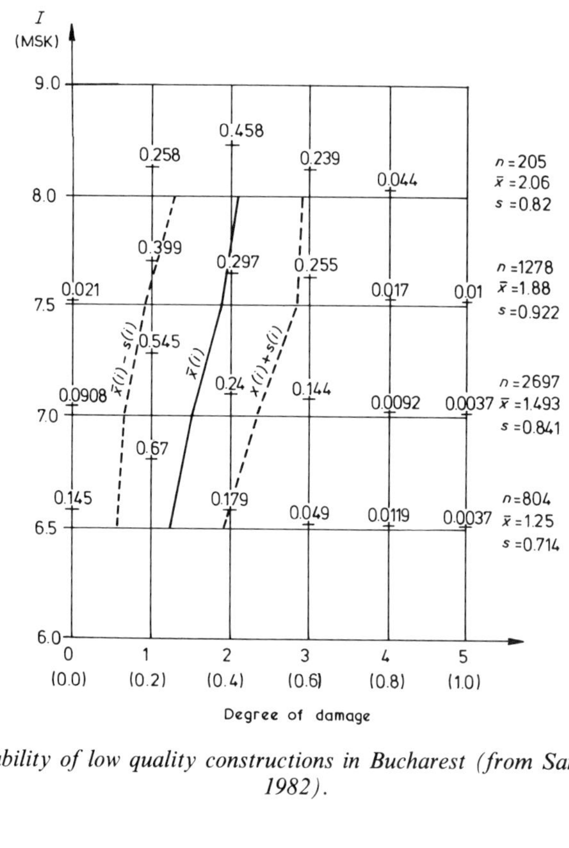

Fig. 1 *Vulnerability of low quality constructions in Bucharest (from Sandi and Vasilescu, 1982).*

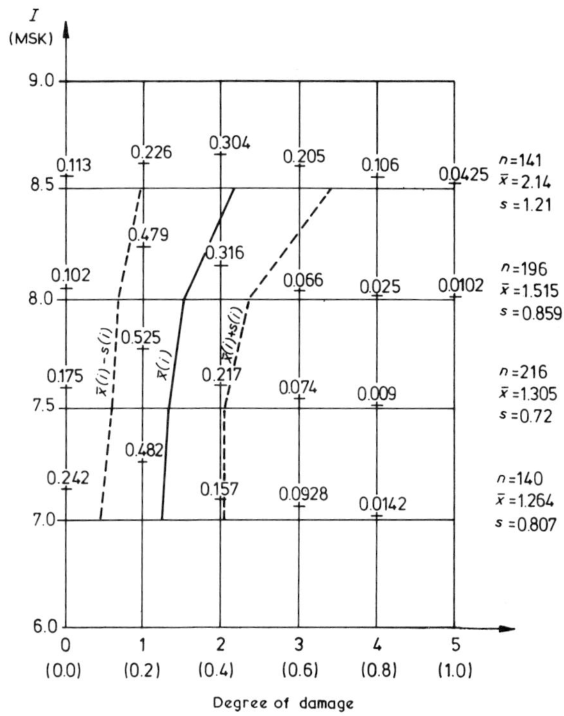

Fig. 2 *Vulnerability of reinforced concrete frame structures in Bucharest (from Sandi and Vasilescu, 1982).*

3 Derivation of vulnerability functions from the observed damage

The general survey of damage in Bucharest after the March 1977 earthquake revealed that the MSK intensity varied between 6.5 and 8.5 from one part of the urban area to another, depending presumably on local soil conditions. This made it possible to derive empirical vulnerability functions within that range of intensities for various types of building.

The results for two building types (low quality constructions and reinforced concrete frame structures) are shown in Figs 1 and 2. These figures show, for each intensity zone, the proportion of buildings of each type suffering the various degrees of damage from 0 to 5. The number n of such buildings in each zone, the mean value $\bar{x}$ of the degree of damage and its standard deviation s in each intensity zone are shown on the right-hand side of the figures. The solid line $\bar{x}(i)$ shows the variation in the mean value of the degree of damage with intensity and the broken lines show the positive and negative standard deviations from the mean. Figure 3 shows similar vulnerability diagrams for one of the types of buildings damaged by the April 1979

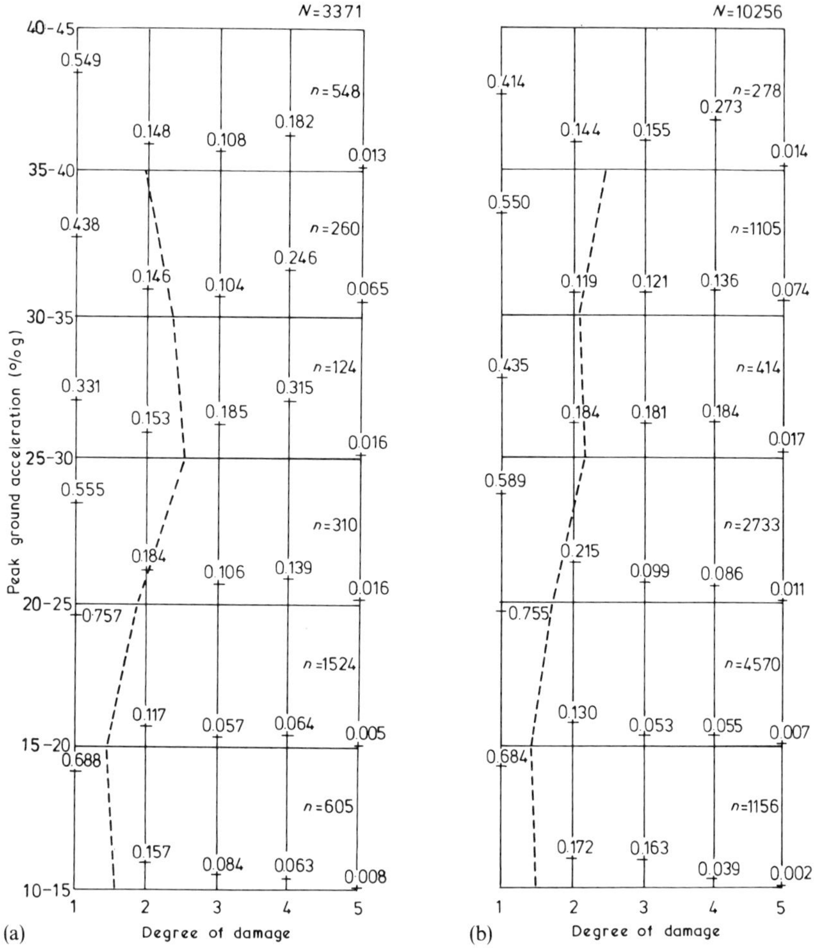

Fig. 3 *Vulnerability of strengthened masonry buildings in Montenegro: (a) on rock; (b) on alluvial ground.*

earthquake in Montenegro, subdivided according to whether the buildings were situated on rock or on alluvial ground. In this case, six intensity zones are defined by ranges in the values of the estimated peak ground acceleration on "firm ground".

The most notable feature of Figs 1–3 is the wide scatter in degree of damage at all levels of intensity and for all building types. This scatter may be due to variations in dynamic properties from one building to another within a given construction type and/or to variations in ground motion intensity from site to site within each intensity zone. Since the variations in ground motion intensity from one part of the urban area of Bucharest to another over distances of the order of 1 km were taken into account in the analysis, the scatter in degree of damage observed in Figs 1 and 2 is presumably due mainly to variations in the dynamic properties of buildings within each construction type. The scatter in degree of damage is even more marked in Fig. 3, and here it is probably also due to local departures from the estimated values of peak ground acceleration which, as we have seen, are based on interpolation and extrapolation, over quite large areas, of data obtained at only six points within the macroseismic area.

4 *The assessment of specific risk*

The method of estimating specific risk adopted here is similar to that suggested by Sandi (1982). We assume that the buildings examined after the earthquakes in Romania and Yugoslavia were typical of all such buildings of the same type in so far as their seismic vulnerability is concerned. If this assumption is correct, the observed distributions of degree of damage at each intensity level, as seen in Figs 1–3, can be taken as representing the probability distributions of degree of damage in any future event of that intensity. We can then read off from these figures the values of p_{id}, i.e. the probability of a degree of damage $D = d$ to any building of a particular type in the event of ground motion of intensity $I = i$ at the site.

The annual probability p_d that a degree of damage d due to all possible events will occur is .

$$p_d = \sum_{i=0}^{i=i\ \max} p_i p_{id}$$

where p_i is the annual probability that ground motion of intensity i will occur. The annual probabilities of specific loss p_l corresponding to p_d can then be computed using the relations given in Table 3 or, if they are available, other relations based on more detailed knowledge of repair costs.

4.1 The case of Bucharest

Studies of the seismicity of Romania have yielded estimates of the annual probabilities of occurrence of seismic ground motion of various intensities in Bucharest (Radu *et al.*, 1982). These estimates are reproduced in Table 4 together with our interpolations to give values of p_i for half-degree intensity intervals.

The results given in Table 4 are derived from the analysis of data on past earthquakes, for each of which a single value of intensity was adopted for Bucharest as a whole; the value adopted for the earthquake of 4 March 1977 was 8. However, the intensity in various parts of the urban area during this earthquake appears to have ranged from 6.5 to 8.5, presumably because of variations in local soil conditions. If these variations are confirmed by microzoning, the values of p_i in Table 4, which are valid for the zone of intensity 8 in the March 1977 earthquake, will have to be adjusted as appropriate for the other zones. This can be done simply

Table 4 *Annual probabilities p_i of seismic intensities in Bucharest*

Intensity (MSK)	6	6.5	7	7.5	8	8.5	9
p_i(Radu *et al.*, 1982)	0.099		0.035		0.014		0.007
p_i(interpolated)	0.063	0.036	0.022	0.013	0.009	0.005	0.004

by shifting the lowest row of values to the left or right, so that the value 0.009 corresponds to the intensity observed in March 1977.

We can now proceed to calculate the specific risk (expressed as an annual average specific loss) for reinforced concrete frame structures in the zone of intensity 8 (March 1977) in Bucharest, using the values of p_{id} taken from Fig. 2, the values of p_i taken from Table 4 and the values of specific loss l_d corresponding to each degree of damage d shown in Table 3. The results are shown in Table 5. It should be noted that in this calculation we have used only the available data on damage, which were limited to buildings in intensity zones ranging from 7 to 8.5. For a more precise estimate of specific risk we should include the contribution of possible damage due to intensities both below and above this range, but this must of course await the accumulation of further observational data.

Using the appropriate values of p_i, we can calculate in a similar way the specific risk to reinforced concrete frame structures in the other intensity zones of the Bucharest urban area. The results are shown in Table 6 together with similar figures for low quality constructions, for which the data on vulnerability are taken from Fig. 1. The interesting feature of Table 6, rather than the actual values of specific risk which are subject to all the sources of error mentioned above, is the variation in specific risk from one intensity zone to another. The specific risk to low quality buildings in zone 8 is higher by a factor of 2.51 than it is in zone 7; in the case of reinforced concrete frame structures, the factor is 2.35. Similar values of this factor are found for other building types.

The application of codes of earthquake-resistant design and construction

Table 5 *Calculation of specific risk for reinforced concrete frame structures in Bucharest (intensity zone 8)*

Intensity	p_i	$p_i p_{id}$				
		$d = 1$	$d = 2$	$d = 3$	$d = 4$	$d = 5$
7	0.022	0.0108	0.0035	0.0020	0.0003	—
7.5	0.013	0.0069	0.0028	0.0010	0.0001	—
8	0.009	0.0039	0.0026	0.0005	0.0002	0.0001
8.5	0.005	0.0013	0.0017	0.0012	0.0006	0.0002
p_d		0.0229	0.0106	0.0047	0.0012	0.0003
$p_l = p_d l_d$		0.0009	0.0017	0.0017	0.0008	0.0003

$r = \Sigma p_l = 0.0054 = 0.54\%$

Table 6 *Specific risk to low quality buildings and reinforced concrete frame structures in various intensity zones in Bucharest*

Intensity zone (March 1977)	*Risk (%)*	
	Low quality buildings	*Reinforced concrete frame structures*
6.5	0.25	—
7	0.39	0.23
7.5	0.60	0.35
8	0.98	0.54
8.5	—	0.84

should tend to reduce this factor, but for non-engineered buildings the specific risk appears to increase, in Bucharest at least, by a factor of about 2.5 for each increase of 1 degree (MSK) in intensity. This has obvious implications for urban planning in seismic zones.

The value of this factor depends, of course, on the slope of the intensity–frequency curve for the area in question. This in turn depends on the *b* value in the magnitude–frequency relation of earthquakes in the nearby seismogenic zones. The greater the *b* value is, the greater will be the variation in the level of risk between one intensity zone and another.

4.2 The case of Montenegro

The seismic hazard in the southern coastal zone of Montenegro can be represented by the estimates given in Table 7 for the city of Bar, made by the Republic Institute of Seismology, Titograd (personal communication). The procedure for calculating the

Table 7 *Seismic hazard in the southern coastal zone of Montenegro*

Peak ground acceleration (%*g*)	10–15	15–20	20–25	25–30	30–35	35–40
Annual probability p_i	0.099	0.036	0.013	0.007	0.004	0.002

specific risk when the intensity is expressed in terms of acceleration is similar to that adopted for the case of Bucharest. We can take as an example the case of strengthened masonry buildings situated on alluvial ground (Table 8).

A feature of Table 8, which also appears in the tables of calculation for other building types in both Romania and Yugoslavia, is the high proportion of risk contributed by the lower but more frequent intensities of ground motion. In the Montenegro case, the specific risk due to ground motion with acceleration in the range 10–15%*g* contributes 0.65% to the total (almost exactly half). The risk

Table 8 *Specific risk to strengthened masonry buildings in the coastal zone of Montenegro*

Acceleration (%g)	p_i	$p_i p_{id}$			
		d = 2	*d* = 3	*d* = 4	*d* = 5
10–15	0.099	0.0170	0.0161	0.0039	0.0002
15–20	0.036	0.0047	0.0021	0.0020	0.0003
20–25	0.013	0.0028	0.0013	0.0011	0.0001
25–30	0.007	0.0013	0.0013	0.0013	0.0001
30–35	0.004	0.0005	0.0005	0.0005	0.0003
35–40	0.002	0.0003	0.0003	0.0005	—
p_d		0.0266	0.0216	0.0093	0.0010
$p_l = l_d p_d$		0.0017	0.0054	0.0052	0.0010

$r = \Sigma p_l = 0.0133 = 1.33\%$

corresponding to intensity 7 for reinforced concrete frame structures in Bucharest represents about one-third of the total (see Table 5). The specific risks to other types of buildings in the southern coastal zone of Montenegro, calculated using the same method, are shown in Table 9. No estimates of specific risk are given in Table 9 when the number of buildings inspected was less than 1000.

4.3 Discussion

The estimates of specific risk (average annual specific loss) given above are provided as examples of a method of calculation. Because of the probable errors inherent in this method, they must be treated with great caution. The main source of error and uncertainty in the assessment of seismic risk arises from our inability (as yet) to find a satisfactory solution to the following problems: how, and by what parameter, can we express the intensity of seismic ground motion, and how can we express seismic hazard in terms of that parameter?

These problems present formidable difficulties, firstly because the damaging power of ground motion depends not only on the peak acceleration but also on many other factors, such as the frequency spectrum, the duration of shaking etc., and secondly because little is yet known about variations in the attenuation of seismic

Table 9 *Specific risks in the coastal zone of Montenegro*

Type of building	*Specific risk* (%)	
	Rock	*Alluvium*
Stone masonry	5.92	6.18
Brick masonry	—	3.30
Strengthened masonry	1.51	1.33
Reinforced concrete frame structure	—	1.18

waves from one region to another or about the precise effects of local topography and soil conditions. Since no easy solutions to these problems are in sight, it may be worth examining other possible approaches to the assessment of seismic risk.

One such approach might be to ignore entirely the science of seismology and to regard the occurrence of damaging ground motion as an unexplained phenomenon to be analysed on a purely statistical basis using the common macroseismic scales of intensity. This approach has the advantage of simplicity, and a more careful analysis of damage in future earthquakes could possibly lead to the adoption of separate "intensity" scales for different types of buildings and to different estimates of hazard and risk for each type.

An alternative approach would be to relate damage directly to the magnitude and epicentral distance of the earthquake, i.e. to the fundamental seismological parameters. This approach also has the advantage of simplicity, but the disadvantage that any empirical relation between degree of damage, magnitude and distance will only be of local validity because of the uncertainties regarding the variation in attenuation from region to region. It may nevertheless be of interest to apply this approach to the April 1979 earthquake in Montenegro.

5 Relation between degree of damage and epicentral distance

Figure 4 shows the observed distributions of degree of damage (from 1 (no damage) to 5 (collapse)) during the April 1979 earthquake in Montenegro for strengthened masonry buildings located on rock and on alluvial ground at various distances from the epicentre. The broken lines link the mean values of degree of damage in each distance range. In the absence of appropriate data on earthquake source zones and magnitude–frequency relationships, we cannot use the data in Fig. 4 to calculate specific risk. However, it can be seen by comparing Fig. 4 with Fig. 3 that the changes in the shape of the damage distribution on moving away from the epicentre (Fig. 4) resemble quite closely the changes in shape obtained on moving from higher to lower ranges of estimated peak ground acceleration (Fig. 3). Of course, this is not surprising because the epicentral distance is a significant factor in inferring the areal distribution of peak ground acceleration from a limited number of accelerograph records.

Figure 5 shows the mean values of observed degree of damage for three types of buildings in Montenegro as functions of both epicentral distance and estimated peak ground acceleration. These functions may be of use in assessing probable degrees of damage and loss in future earthquakes in the same area. However, they cannot be used to estimate seismic risk to such buildings in any other area without detailed knowledge of the way in which attenuation varies from one area to another.

By relating damage directly to epicentral distance it may be possible to avoid the difficulty of choosing a parameter to express the intensity of seismic ground motion, but there still remains the problem of attenuation and its probable variations from region to region. Progress in seismic risk assessment must therefore

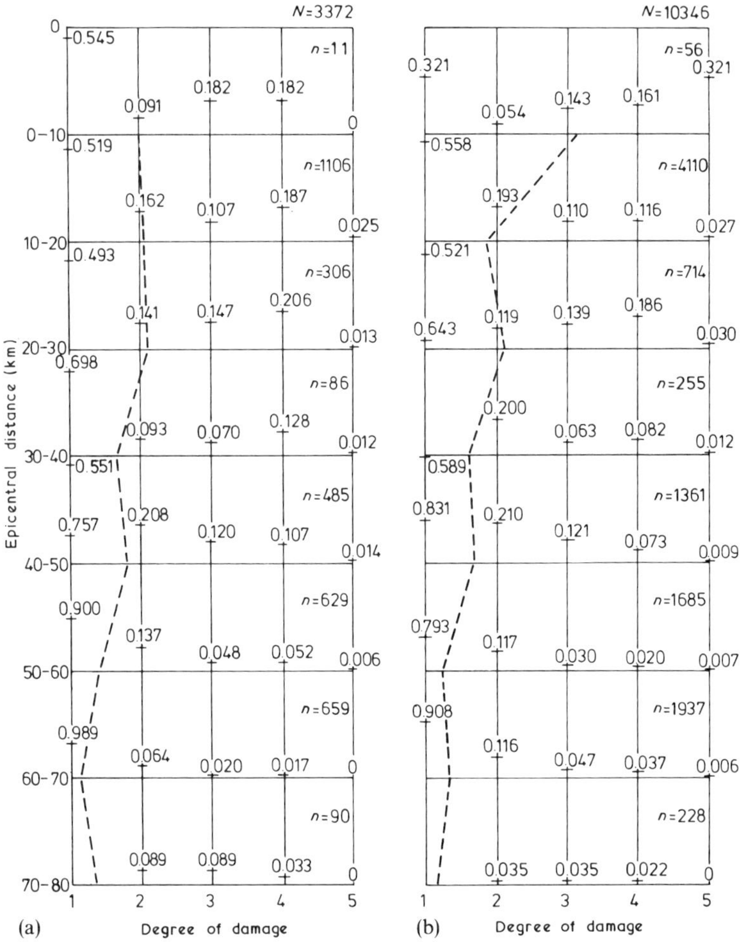

Fig. 4 Degree of damage versus epicentral distance for strengthened masonry buildings in Montenegro: (a) on rock; (b) on alluvial ground.

await the accumulation of a much greater volume of observational data on both strong ground motion and damage to buildings. Strong-motion accelerographs are being installed in rapidly increasing numbers. Efforts must also be made to use systematic damage surveys such as those already undertaken in the Balkan countries to extract the maximum possible amount of useful information from buildings damaged in any future earthquake.

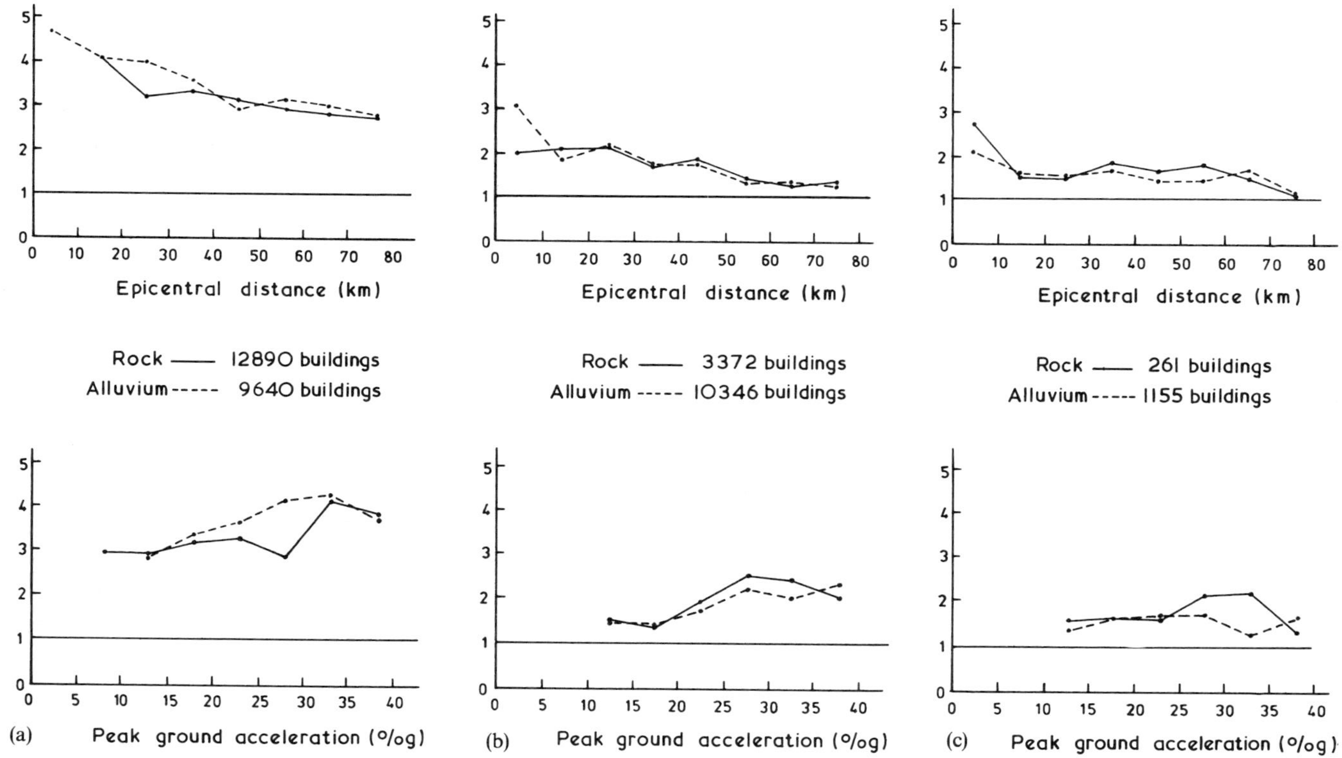

Fig. 5 *Average degree of damage to three types of buildings in Montenegro during the earthquake of 15 April 1979 as functions of the epicentral distance and the estimated peak ground acceleration: (a) stone masonry; (b) strengthened masonry; (c) reinforced concrete frame structures.*

Acknowledgments

I am indebted to Professor Jakim Petrovski and the Director and staff of the Institute of Earthquake Engineering and Engineering Seismology, University of Skopje, for their kindness in making available the data on the damage sustained by buildings in Montenegro during the earthquake of 15 April 1979.

References

PETROVSKI J., NOCEVSKI N., RISTIC D., MILUTINOVIC Z., STANKOVIC V. and PAVICEVIC B. (1984): An applied methodology for the assessment of earthquake vulnerability and risk, based on the earthquake of 15 April 1979 in S. R. Montenegro (SFR Yugoslavia), *Report No. 84–084*, Institute of Earthquake Engineering and Engineering Seismology (IZIIS), Skopje, December.

RADU C., UTALE A. and ARDELEANU L. (1982): National report of Romania. In *Final Report of Working Group A (Seismology, Seismotectonics, Seismic Hazard and Earthquake Prediction), UNDP Project RER/79/014*, UNESCO, Athens, December.

SANDI H. (1982): Mathematical concepts and relations concerning vulnerability and risk, *UNDP Project RER/79/014 Working Group C: Final Report, Annex A2*, UNESCO, Sofia.

SANDI H. and VASILESCU D. (1982): National report of Romania on observed vulnerability of buildings. In *Final Report of Working Group B (Vulnerability and Seismic Hazard), UNDP Project RER/79/014*, UNESCO, Skopje, December.

UNDRO (1979): *Natural Disasters and Vulnerability Analysis: Report of Expert Group Meeting*, United Nations, Geneva.

Estimation of Expected Seismic Vulnerability: A Simple Methodology for Medium-sized Groups of Older Buildings

PETER SHEPPARD AND MARJANA LUTMAN

Institute for Research and Testing in Materials and Structures, Dimičeva 12, 61000 Ljubljana, Yugoslavia

Abstract A simple methodology for the estimation of the seismic vulnerability and relative seismic risk of older predominantly masonry buildings is described. The methodology is suited to medium-sized groups of heterogeneous buildings of the type often found in urban planning studies. It aims, through on-site inspections and an assessment of the most important structural characteristics of the buildings, to provide meaningful comparisons of their vulnerability. By taking into account the relative density of users within groups of buildings serving a similar purpose, the relative seismic risk associated with each building can be determined. The results of such studies, two of which are presented here, can be used as a basis for deciding on the priority to be given to individual buildings with respect to aseismic strengthening measures.

1 Introduction

In many countries the seismic resistance of a considerable proportion of all buildings is insufficient to meet the requirements of the official technical regulations. Such buildings are often simply referred to as "older buildings". In Yugoslavia this definition refers to buildings whose construction was completed prior to the introduction of the first national aseismic regulations in 1964, a year after the Skopje earthquake. The estimation of the seismic vulnerability of older buildings is an important task for engineers and planners in both pre-earthquake and post-earthquake periods.

Methods for assessing the seismic vulnerability (or seismic resistance) of older predominantly masonry buildings have developed mainly as the consequence of the severe damage and collapse of these buildings in past earthquakes, and new regulations, based on these assessments, have been introduced in order to avoid such disasters in the future. In Yugoslavia, such scientific studies began after the Skopje earthquake of 1963 when numerous multistorey brick masonry buildings collapsed or were badly damaged. These studies have become more intensive following the numerous strong earthquakes which have struck different parts of Yugoslavia in the last two decades. The earthquake damage to buildings has been recorded by on-site investigation teams, and this information has been used to help verify assumed damage mechanisms and to confirm the adequacy of analytical seismic resistance calculations.

An early attempt to create an up-to-date database concerning damage caused to buildings and the possible costs of repairs and strengthening was made after the Kozjansko earthquake of June 1974 in eastern Slovenia, a republic situated in the northwest part of Yugoslavia. The methods used at that time were later improved to deal with the consequences in western Slovenia of the Friuli earthquakes of May and September 1976. Both the Kozjansko and Friuli earthquakes affected less developed parts of Slovenia, where stone masonry buildings one to three storeys high are the most common. The collected data concerning the damage to older buildings caused by these earthquakes were used, together with the results of laboratory investigations of the behaviour of wall elements and model buildings, to verify the adequacy of methods for quantifying the seismic resistance of stone masonry buildings (Turnšek *et al.*, 1978).

Many important results were obtained from an analysis of the consequences of the great Montenegro earthquake of 1979, which caused damage to more than 40 000 buildings. In this case it was possible to carry out large-scale statistical analyses of the performance of different types of buildings, including those built of unstrengthened stone masonry and unstrengthened brick masonry (i.e. masonry buildings without horizontal and/or vertical reinforced concrete tie-beams). The results of these studies have been presented in the literature (Petrovski *et al.*, 1984) in the form of diagrams which relate the observed vulnerability, expressed as the lost percentage of the gross area of the type of building under consideration, to the equivalent ground acceleration at 300 different locations (settlements). This acceleration was derived from recorded ground motion, taking into account the local soil conditions for each settlement.

An official methodology for the assessment of damage due to natural hazards in Yugoslavia was adopted by law in 1979 and improved upon in 1987. However, no all-embracing methodology yet exists to determine the pre-earthquake "expected" or "predicted" vulnerability of older buildings with insufficient seismic resistance. This is particularly true for medium-sized groups of such buildings, for example in urban renewal studies, where appropriate analytical and experimental studies of individual buildings cannot be economically justified and yet the buildings vary too much among themselves to be grouped together as typical representatives of widely differing types of buildings.

Thus the purpose of this paper is to present a simple methodology for the estimation of the seismic vulnerability of medium-sized groups of heterogeneous older buildings which takes into account, on the basis of on-site inspections (which are not, however, accompanied by experimental or analytical investigations), the most important structural parameters influencing the seismic resistance of these buildings. The results of an application of this methodology to two differing groups of older predominantly masonry buildings are presented. These studies concern the vulnerability of, in the first case, 27 buildings forming a small planning zone in the old part of the city of Ljubljana and, in the second case, 22 older buildings belonging to elementary, secondary and other schools in one of the communes of the city of Ljubljana.

2 *Development of a simple methodology*

2.1 Brief literature survey

When developing a suitable simple methodology for the estimation of the expected seismic vulnerability of heterogeneous older buildings, existing methodologies presented in the literature and our own experience were both taken into account. It follows from the literature (a general survey of which has been provided by Corsanego (1985)) that a large number of different methodologies exist, most of which are, to a greater or lesser extent, suited to the specific conditions in particular countries. The choice of a methodology suitable for use in a particular area depends mainly on three factors: the size and nature of the stock of buildings which is to be examined, the objectives of the analysis (e.g. for urban planning, civil defence purposes or the preparation of a long-term strategy for the reduction of seismic risk) and the funds and time available for the study to be carried out.

Only those methodologies which are based on the principles of categorization of buildings and their structural elements are considered in this paper. Categorization is carried out mainly on the basis of experience regarding the behaviour of buildings in past earthquakes. In this case the expected seismic vulnerability of buildings is estimated by means of *in situ* inspections of individual buildings in which the earthquake-resistant properties of their structural elements and of the building as a whole are assessed. Any available technical documentation for the building is inspected, but as a rule no analytical calculations of seismic resistance are made although previously calculated results referring to similar buildings can be used for orientation purposes. In particular, careful note is made of any structural damage which already exists, often in the form of cracks, and which can provide a guide to the building's possible future behaviour under strong seismic loads.

One of the first systematic methods for the estimation of the seismic vulnerability of buildings—the field evaluation method—was presented by Culver *et al.* (1975). This method, which is applicable to buildings of all types, takes into account the following factors, each with several possible different ratings: the type of resisting elements, the symmetry of the resisting elements, the quantity of the resisting elements, the present condition of the resisting elements, the rigidity of the structure, the anchorage and connections, and the presence or absence of chords or ties. A somewhat similar method was proposed by Wiggins *et al.* (1974). This method takes into account the framing system and/or walls, the diaphragm and/or bracing system, the interior partitions, the physical condition of the structure and special hazards. By taking into account a combination of the estimated ratings, as well as the local seismicity or expected ground acceleration, "risk ratings" and "risk safety factors" which indicate the relative seismic risk of individual buildings within a group are defined by these two methodologies. At this point an additional factor concerning the structural or socio-economic importance of individual buildings can be introduced. Another interesting methodology, based on the concept of "de-merit points", has been presented by Shepherd (1978).

In recent years there has been a considerable increase in interest in the question of the seismic vulnerability of older buildings in Italy. Over the last decade, many such buildings, particularly in historical old town centres, have been badly damaged by strong earthquakes. The method proposed by Benedetti and Petrini (1984) has become popular for the estimation of their seismic vulnerability. This method, which is used to calculate the "vulnerability index" of individual buildings, takes into account the following 10 parameters, each of which is assigned a weighting factor varying from 0.25 to 1.5: the connection of the walls, the type of wall, the soil conditions, the total shear resistance of the walls, plan regularity, elevation regularity, the horizontal diaphragms, the roof details and general maintenance conditions. According to each of these parameters, buildings can be placed in one of four classes, denoted A, B, C and D. Class A represents conditions in accordance with the valid aseismic regulations and Class D represents "unsafe conditions". The vulnerability index of a building is given by the sum of the numerical values corresponding to each parameter, each multiplied by the weighting factor given for that particular parameter.

An opportunity for evaluating the suitability of the Benedetti–Petrini method occurred in 1984, when an earthquake of intensity VII on the MCS scale struck the town of Gubbio in central Italy. Shortly before this earthquake, the method had been used to assess the seismic vulnerability of the buildings in the town's historical centre. Some details of the correlation obtained between the observed and predicted degree of damage at Gubbio (approximately 800 buildings) have been given by Benedetti and Benzoni (1985), who compared these results with those obtained after the Friuli earthquakes of 1976 (a total of approximately 1500 buildings were included in this case). They found fairly satisfactory agreement, although they thought that some corrections of the scoring and weighting factors might be necessary. At the greater seismic intensities observed in Friuli (VIII to X on the MCS scale) there was better correlation between the total damage observed and the parameter "total shear resistance of the walls", whereas at lesser intensities there was better correlation with the parameter "horizontal diaphragms".

2.2 The parameters chosen for the methodology

On the basis of comparisons between the advantages and disadvantages of the methodologies described above, in the case of masonry buildings it was decided to take into account only five main parameters of seismic vulnerability, with the following weighting factors w:

(1) type and quality of walls, $w = 1.0$;
(2) quantity of walls (relative to the floor plan), $w = 1.0$;
(3) layout of walls in the floor plan, $w = 1.0$;
(4) connectedness of the walls, $w = 1.5$;
(5) other factors (condition of building etc.), $w = 0.5$.

In each case an individual parameter is given an integral value ranging from 1 to 5, where a value of 1 means conditions in accordance with the valid technical regulations and a value of 5 means conditions which do not correspond at all to the regulations. The sum of these values, each multiplied by the given weighting factor, is called the basic seismic vulnerability P_s of the building, and it can assume values between 5.0 and 25.0.

The first three parameters mentioned above are the same as those taken into account by Turnšek *et al.* (1982) in a parametric analysis of the seismic resistance of masonry buildings. This method involves a step-by-step estimate of the base shear coefficient at failure and requires specific data on building height, wall distribution in both directions, characteristic wall strength and vertical loading. In the case of the methodology presented here, however, building height is taken into account by means of an additional parameter P_h which assumes the following values:

(1) buildings up to three storeys high (ground plus two) of normal storey height, $P_h = 0$;

(2) buildings three storeys high with above-average storey height, and buildings four storeys high with normal storey height, $P_h = 2$;

(3) buildings four storeys high with above-average storey height, and buildings five storeys high with normal storey height, $P_h = 4$;

(4) buildings five storeys high with above-average storey height, and buildings six storeys high with normal storey height, $P_h = 6$.

Buildings with above-average storey height are considered to be those whose average storey height exceeds 3.5 m.

Apart from ordinary masonry buildings, numerous pre-1964 buildings whose load-bearing systems consist of a mixture of masonry walls and reinforced concrete columns exist in Yugoslavia. The following somewhat altered basic parameters of seismic vulnerability have been chosen for these buildings (the distribution of the weighting factors has been kept the same):

(1) type and quality of walls and columns, $w = 1.0$;
(2) quantity of walls and columns (relative to the floor plan), $w = 1.0$;
(3) layout of walls and columns in the floor plan, $w = 1.0$;
(4) mix of the structure and structural details, $w = 1.5$;
(5) other factors (condition of building etc.), $w = 0.5$.

Since Slovenia has seismic zones with a maximum expected intensity of IX on the MCS scale, the methodology presented here takes into account only local seismicity of degree VIII and IX. In order to be able to compare the relative seismic risk to buildings in different zones, an additional parameter P_1 relating to general seismicity and to local soil conditions, which makes use of the seismic coefficients prescribed for microzoning purposes, has been introduced. The values adopted are given in Table 1.

Table 1

Zone	*Seismic coefficient* K_s	P_1
VIII–1	0.04	−3
VIII–2	0.05	0
VIII–3	0.06	+3
IX–1	0.08	+9
IX–2	0.10	+15
IX–3	0.12	+21

For all ground conditions the seismic coefficients for a seismic zone of expected maximum intensity IX are twice the size of those for a seismic zone of intensity VIII. Since the average value of the basic seismic vulnerability of buildings in different zones is $P_s = 15$ points, the same number of points has been allocated to the span from VIII–2 to IX–2 (ground of medium quality). Although this approach could be considered somewhat questionable, it is a simple and efficient way of taking into account the ground conditions, which are often fairly uncertain, at different locations, particularly within an area of uniform expected seismic intensity.

When all the parameters mentioned above are taken into account, an expression can be written defining the seismic vulnerability of older buildings:

$$V_s = (P_s + P_h + P_1)F_t \tag{1}$$

where F_t is a factor for the type of load-bearing system, whose value is taken to be 1.0 for masonry buildings and 0.85 for buildings with masonry walls and reinforced concrete columns (i.e. a general 15% reduction in the vulnerability of these buildings compared with masonry buildings due to the somewhat greater ductility of the former has been taken into account).

Whereas an estimate of the seismic vulnerability of a building refers mainly to the damage which may be caused by an earthquake of a certain intensity, the expression seismic risk usually refers to the users of a building and, to a certain extent, the value of the building and its contents. For estimates of seismic risk, which are of particular importance when deciding the priority of individual buildings within renewal programmes, the two most important parameters are (a) the number of users of the building and (b) the total usable floor area of the building. The average density of users of the building can be determined from these two parameters. Of course the number of users will change during the day and during the year, but for simple analyses it is sufficient to use an average density which is as realistic as possible. First the density of users in individual buildings is determined (e.g. the number of users per 100 m^2 of usable floor space) and then this figure is compared with the weighted average density of users for all buildings of comparable purpose. In this way the relative density D_r of users within a group of buildings can be determined. The relative seismic risk which can be allocated to an individual building is then given by the expression

$$V_u = V_s D_r \tag{2}$$

Since the relative density of users applies to a particular group of buildings having a similar purpose (e.g. residential buildings, commercial buildings), valid comparisons of seismic risk can only be made within that group of buildings. Therefore, for the purposes of this paper, reference is made only to “relative seismic risk”. A somewhat similar approach could be made respecting the actual value of buildings and their contents. Since such evaluations are subject to broad discussion, however, particularly in the case of historic buildings, this question is not treated further here.

3 *Results of application of the methodology*

3.1 Planning zone in the old part of Ljubljana

The methodology described above was applied to a small planning zone in the old city centre of Ljubljana (Fig. 1). This area of about 250 m × 50 m contains a total of 30 houses which serve residential, commercial or mixed residential–commercial purposes. Three of the buildings were already undergoing complete renewal and aseismic strengthening, leaving a total of 27 older buildings, of which five were built after the Ljubljana earthquake of 1895 and the remainder were mostly several hundred years old. The load-bearing walls of the older buildings are stone masonry or mixed stone and brick masonry, whereas those built after 1895 have brick masonry walls. The floor structures are predominantly wooden, with some brick vaults in the lower storeys and, in a few cases, reinforced concrete floor slabs. Most of these buildings are three or four storeys high and have fairly irregular floor plans. In past centuries they have been frequently subjected to structural alterations and sometimes increased in height. Very little relevant technical documentation exists. The planning zone is situated in an area where earthquakes of intensity up to VIII on the MCS scale are expected, with poor ground conditions (VIII–3: $K_s = 0.06$, $p_l = +3$).

Fig. 1 *Typical buildings in the planning zone studied in the old part of the city of Ljubljana.*

The buildings in the planning zone were divided into three types: residential buildings, where more than 75% of the total usable floor area is used for residential purposes, commercial buildings, where more than 75% of the total usable floor area is used for commercial and trade purposes, and mixed-purpose buildings, which form the remainder. In this way three groups of buildings have been obtained consisting of six residential buildings (R1–R6), 13 mixed-purpose buildings (M1–M13) and eight commercial buildings (C1–C8). The weighted average density of users in these three groups of buildings differs considerably, and amounted at the time of the survey to 2.62, 3.59 and 4.84 users respectively per 100 m^2 of usable floor space.

The basic data for the buildings investigated and the seismic vulnerability and relative seismic risk associated with them, as determined by on-site surveys, are shown in Table 2. The seismic vulnerability of individual buildings is shown graphically in Fig. 2, and the relative seismic risk associated with each building is compared with the evaluated vulnerability in Fig. 3 (the order in which the buildings appear is now changed). Figure 4 shows the relative seismic risk associated with the buildings in comparison with their size, expressed as the total usable floor space in square metres. Figure 5 shows a map of the planning zone, indicating how the relative seismic risk associated with individual buildings has been classified into four different grades (very high, $V_u \geqslant 25.0$; high, $17.5 \leqslant V_u < 25.0$; medium, $10.0 \leqslant V_u < 17.5$; low, $V_u < 10.0$).

Table 2 *Results of the seismic vulnerability study of old city centre buildings in Ljubljana*

Building	*Age*[a]	*No. of storeys*	*Usable floor area* (m^2)	*No. of users*	*No. of users per 100 m²*	*Average density of users in groups*	D_r	V_s	V_u
R1	2	4	2751	74	2.69	2.62	1.03	21.0	21.6
R2	1	4	474	10	2.11		0.81	25.0	20.3
R3	1	3	303	4	1.32		0.50	21.5	10.8
R4	1	4	723	14	1.94		0.74	20.5	15.2
R5	1	4	258	15	5.81		2.22	19.0	42.2
R6	1	3	188	6	3.19		1.22	22.5	27.5
M1	1	3	545	26	4.77	3.59	1.33	17.0	22.6
M2	2	3	248	9	3.63		1.01	19.5	19.7
M3	1	3	345	3	0.87		1.00[b]	18.0	18.0
M4	1	4	1123	43	3.83		1.07	23.0	24.6
M5	1	4	863	34	3.94		1.10	22.0	24.2
M6	1	4	1344	54	4.02		1.12	22.5	25.2
M7	1	4	1024	21	2.05		0.57	19.0	10.8
M8	1	4	758	23	3.03		0.84	19.0	16.0
M9	1	5	2257	108	4.78		1.33	23.5	31.3
M10	2	3	445	5	1.12		0.31	19.0	5.9
M11	2	4	859	32	3.72		1.04	21.0	21.8
M12	1	3	354	10	2.82		0.79	21.0	16.6
M13	1	4	258	6	2.32		0.65	21.5	14.0
C1	1	3	329	8	2.43	4.84	0.50	17.0	8.5
C2	1	2	241	6	2.49		0.51	21.5	11.0
C3	1	3	826	44	5.32		1.10	19.5	21.5
C4	1	3	593	29	4.89		1.01	20.5	20.7
C5	1	4	1030	38	3.69		0.76	17.5	13.3
C6	2	4	1018	23	2.26		0.47	22.0	10.3
C7	1	3	737	30	4.07		0.84	19.0	16.0
C8	1	3	2389	169	7.07		1.46	17.5	25.6

[a] 1, pre-1895; 2, post-1895.
[b] Correction 0.24 → 1.00 owing to variability in the number of users.

3.2 Older school buildings in one of the communes of Ljubljana

The methodology was also applied to 22 buildings used for kindergartens, schools and other educational institutes in Ljubljana-Bežigrad, one of the communes of the city of Ljubljana. A total of four kindergartens, seven buildings used for elementary schools, five buildings used for secondary schools and six other buildings used for educational purposes (university, adult education, special schools) were investigated and were denoted K1–K4, ES1–ES7, SS1–SS5 and F1–F6 respectively. Two of these buildings were built in the nineteenth century, five in the period 1900–1945 and 15 in the period 1945–1964. All these buildings were located in an area where earthquakes

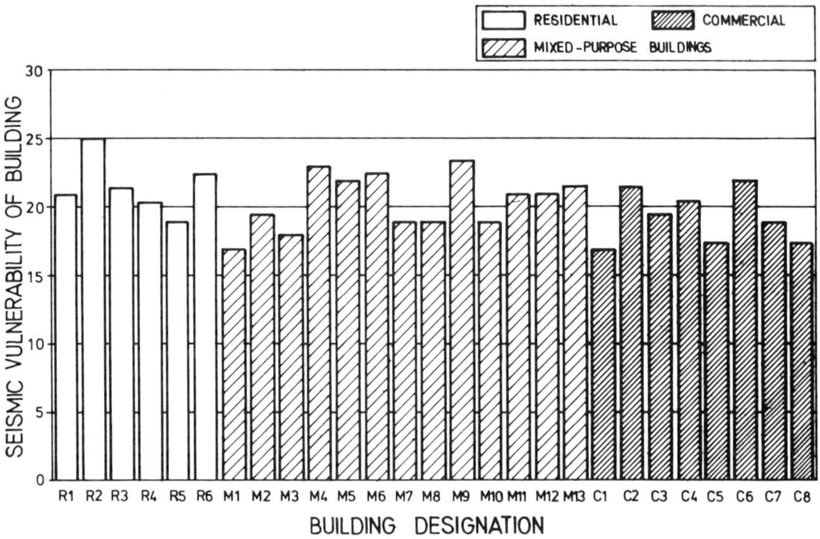

Fig. 2 *Seismic vulnerability of old city centre buildings in Ljubljana.*

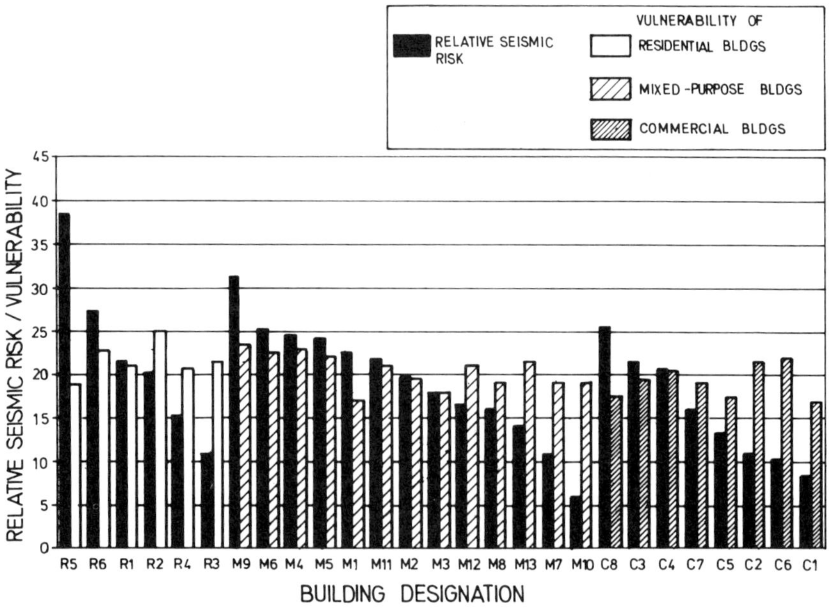

Fig. 3 *Relative seismic risk of old city centre buildings in Ljubljana compared with their seismic vulnerability.*

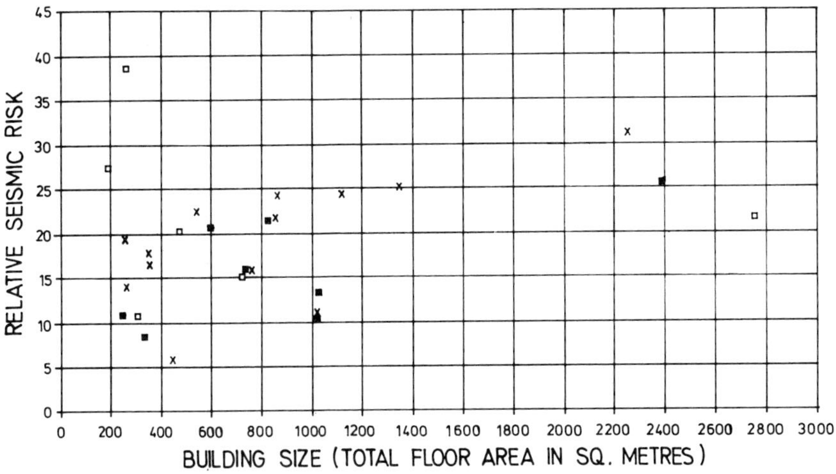

Fig. 4 *Relative seismic risk of old city centre buildings in Ljubljana compared with the building size:* □, *residential buildings;* ×, *mixed-purpose buildings;* ■, *commercial buildings.*

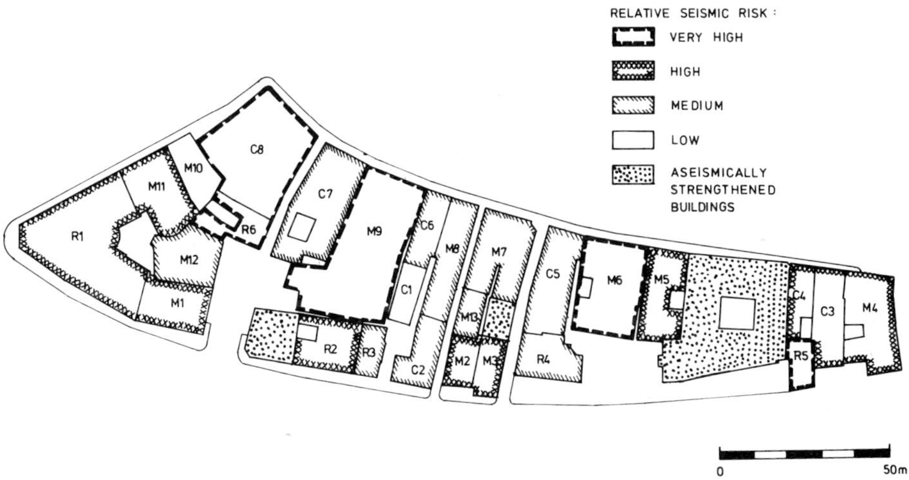

Fig. 5 *Map of the planning zone studied in the old part of the city of Ljubljana, showing the relative seismic risk associated with individual buildings.*

with a maximum intensity of VIII MCS can be expected, with different microzoning seismic coefficients K_s in the range 0.04–0.06. Thirteen of the buildings were masonry structures and nine had load-bearing structures of the mixed type with masonry walls and reinforced concrete columns. Some of the floor structures were wooden and some were reinforced concrete slabs of various kinds. A weighted average of 20 users per 100 m^2 of usable floor space was taken into account.

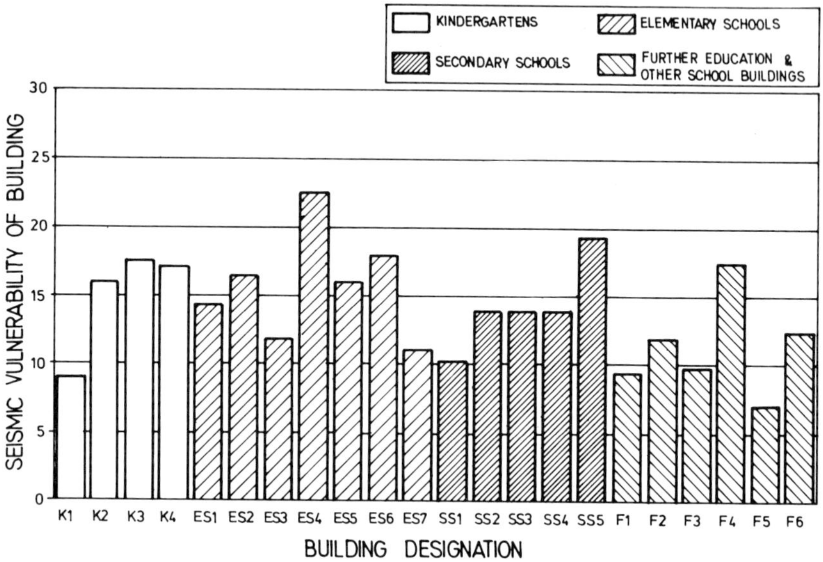

Fig. 6 *Seismic vulnerability of older school buildings in the commune of Ljubljana-Bežigrad.*

The basic data for the buildings investigated are shown in Table 3, and Figs 6, 7 and 8 contain the equivalent results to those shown for the small planning zone in Figs 2, 3 and 4. Comparison of these figures shows that there is a greater variation in expected seismic vulnerability among the school buildings inspected but that the level of vulnerability is in general somewhat lower. As previously, there are large variations in the relative seismic risk associated with individual buildings. The greatest relative seismic risk is associated with two buildings (SS5 and F4) used for adult education (one of which is situated in a former nineteenth century elementary school) and two buildings (K2 and K4) used as kindergartens where a high density of users coincides with seismically weak structures. Building K2 is shown in Fig. 9, and a typical secondary school building (SS2) with a mixed load-bearing structure is shown in Fig. 10.

4 Conclusion

The simple methodology described here for determining seismic vulnerability and relative seismic risk has been found to be satisfactory for comparisons of individual buildings in medium-sized groups of older buildings, i.e. those which do not meet the requirements of modern aseismic codes. Complete analytical and experimental investigations of the seismic strength of buildings in such groups would be too costly, but the buildings vary too much among themselves to be classified into broadly defined structural groups. It is considered that the results of such

Table 3 *Results of the seismic vulnerability study of older school buildings in the commune of Ljubljana-Bežigrad*

Building	*Year built*	*No. of storeys*	*Usable floor area* (m²)	*No. of users*	*No. of users per 100 m²*	*Average density of users*	D_r	V_s	V_u
K1	1938	2	283	104	36.8	20.0	1.84	9.0	16.5
K2	1936	2	389	124	31.9		1.59	16.0	25.5
K3	1948	2	660	115	17.4		0.87	17.5	15.2
K4	1939	2	240	75	31.2		1.56	17.0	26.5
ES1	1958	3	3041	623	20.5	20.0	1.02	14.5	14.8
ES2	1935	4	5956	798	13.4		0.67	16.5	11.1
ES3	1961	3	2993	470	15.7		0.78	11.9	9.3
ES4	1954	1	600	125	20.8		1.04	22.5	23.4
ES5	1902	2	310	60	19.4		0.97	16.0	15.5
ES6	1938	2	708	130	18.4		0.92	18.0	16.5
ES7	1961	2	1157	211	18.2		0.92	11.1	10.2
SS1	1960	3	2866	640	22.3	20.0	1.12	10.2	11.4
SS2	1958	4	3166	800	25.3		1.26	14.0	17.6
SS3	1936	5	5000	970	19.4		0.97	14.0	13.6
SS4	1962	6	4837	800	16.5		0.83	14.0	11.6
SS5	1882	2	465	216	46.5		2.32	19.5	45.2
F1	1961	3	1672	200	12.0	20.0	0.60	9.5	5.7
F2	1961	1	824	246	29.9		1.49	12.0	17.9
F3	1929	2	327	100	30.6		1.53	9.8	15.0
F4	1870	2	302	100	33.1		1.66	17.5	29.0
F5	1964	3	1406	270	19.2		0.96	7.2	6.9
F6	1961	5	2600	800	30.8		1.54	12.5	19.2

comparisons can be used with considerable confidence, particularly in urban renewal studies, to determine the priority to be given to individual buildings with respect to aseismic strengthening measures. Although the methodology intentionally does not include any specific calculations of seismic resistance, the results can be correlated with observed damage to similar buildings in past earthquakes as well as with the results of calculations of seismic strength already performed for buildings of a similar type.

Acknowledgments

The general research into the seismic vulnerability of older buildings was supported by the Research Community of Slovenia and by the Special Research Community for Building and Civil Engineering of Slovenia. The investigations of the vulnerability of buildings in the old part of the city of Ljubljana were supported by the City

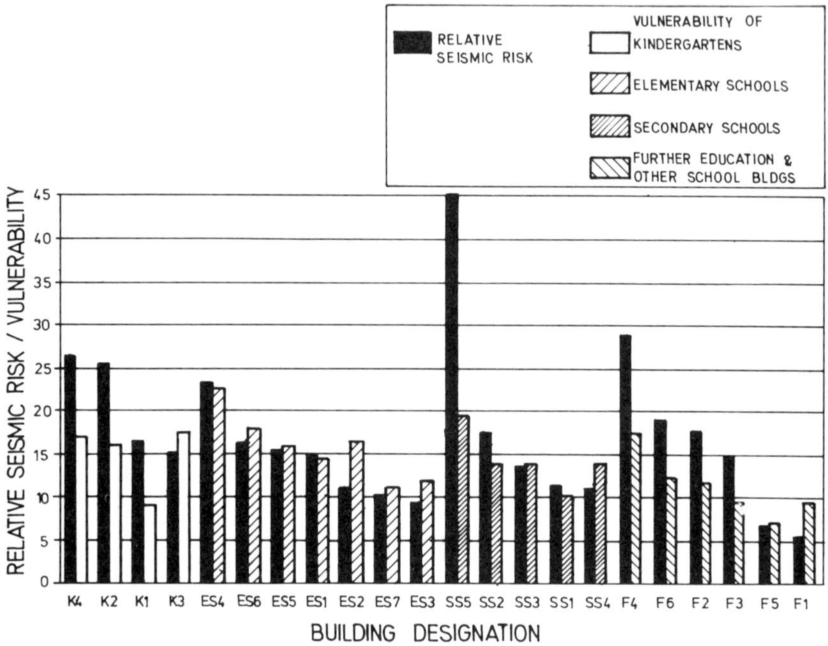

Fig. 7 Relative seismic risk of older school buildings in the commune of Ljubljana-Bežigrad compared with their seismic vulnerability.

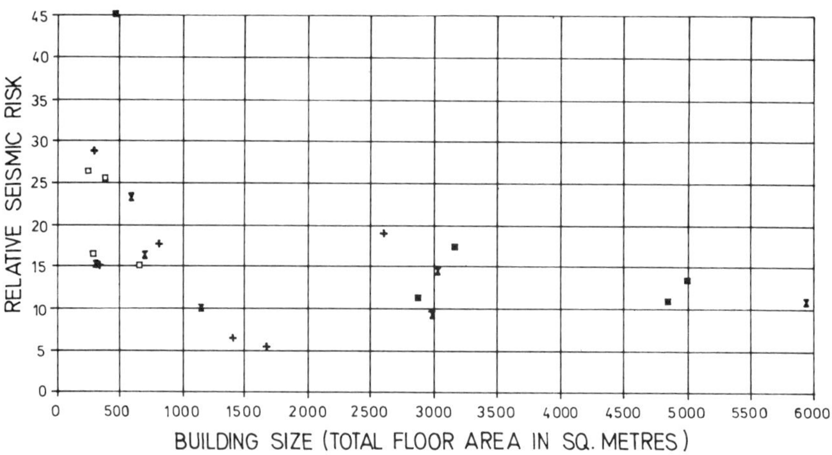

Fig. 8 Relative seismic risk of older school buildings in the commune of Ljubljana-Bežigrad compared with building size: □, kindergartens; ×, primary schools; ■, secondary schools; +, further education and other school buildings.

Fig. 9 *Building K2—example of a building constructed in 1936 used as a kindergarten (a previous out-patients' clinic), with a high density of users (32 per 100 m² of usable floor space) and a seismically weak load-bearing structure (very few internal walls).*

Fig. 10 *Building SS2—example of a secondary school with a load-bearing structure consisting of a mixture of masonry walls and reinforced concrete columns.*

Planning Institute of Ljubljana, and those of the vulnerability of old school buildings were supported by the Research Community of the Commune of Ljubljana-Bežigrad.

References

BENEDETTI D. and BENZONI G. M. (1985): Seismic vulnerability index versus damage for unreinforced masonry buildings, *Proceedings of the International Conference on Reconstruction, Restoration, and Urban Planning of Towns and Regions in Seismic Prone Areas, Skopje, 1985*, pp. 333–347, ANIACAP (Rome), IZIIS (Skopje), CTC (Algiers) and UNESCO (Paris).

BENEDETTI D. and PETRINI V. (1984): On seismic vulnerability of masonry buildings: proposal of an evaluation procedure. *L'Industria delle Costruzioni*, **18** (in Italian).

CORSANEGO A. (1985): A review of methodologies for seismic vulnerability assessment, *Proceedings of the International Conference on Reconstruction, Restoration, and Urban Planning of Towns and Regions in Seismic Prone Areas, Skopje, 1985*, pp. 285–301, ANIACAP (Rome), IZIIS (Skopje), CTC (Algiers), and UNESCO (Paris).

CULVER C. G., LAW H. S., HART G. C. and PINKHAM C. W. (1975): Natural hazards evaluation of existing buildings, *Report NBS BSS-61*, Building Science Series, National Bureau of Standards, Washington, DC.

PETROVSKI J., STANKOVIČ V., NOČEVSKI N. and RISTIČ D. (1984): Development of empirical and theoretical vulnerability and seismic risk models, *Proceedings of the 8th World Conference on Earthquake Engineering, San Francisco, CA*, Vol. 1, pp. 433–440, Prentice-Hall, Englewood Cliffs, NJ.

SHEPHERD R. (1978): High earthquake risk buildings in New Zealand. In *Earthquake Engineering and Structural Dynamics*, Vol. 6, pp. 383–395, Wiley, Chichester.

TURNŠEK V., SHEPPARD P. and TERČELJ S. (1982): Parameters of the seismic resistance of masonry walls and structures, *Proceedings of the British Ceramic Society*, **30**, 370–380.

TURNŠEK V., TERČELJ S., SHEPPARD P. and TOMAŽEVIČ M. (1978): The seismic resistance of stone-masonry walls and buildings, *Proceedings of the 6th European Conference on Earthquake Engineering, Dubrovnik*, Vol. 3, pp. 255–262, Yugoslav Association for Earthquake Engineering, Ljubljana.

WIGGINS J. H., LEE L. T., PLOESSEL M. R. and PETAK W. J. (1974): *Seismic Safety Study: City of Los Angeles*, J. H. Wiggins Company, Redondo Beach, CA.

Intensity Attenuation and Seismic Hazard Analysis

GIUSEPPE GRANDORI

Politecnico di Milano, Piazza Leonardo da Vinci 32, I-20133 Milano, Italy

Abstract Some remarks about the structure of commonly used attenuation laws are presented. The analysis demonstrates the need for greater flexibility in fitting the experimental data. A group of Italian earthquakes is used to define the conditions that the structure of an attenuation law should satisfy. From such conditions a new attenuation law is derived, which shows a very good fit to experimental data. A numerical example shows that the choice of the structure of the attenuation law has a marked influence on the analysis of local seismic hazard.

1 Introduction

When a seismic analysis for a particular site is carried out using historical data the relation between macroseismic intensity and the epicentral distance (the attenuation law) plays a very important role. In this paper it is shown that, even if a large amount of credible statistical data are available for isoseismal maps, large errors can be made in the evaluation of local hazard depending on the structure of the attenuation law that is used to interpret the data.

The problem can be clarified by using an example. The average experimental attenuation of macroseismic intensity derived from isoseismal maps for 19 earthquakes in southern Italy is shown in Fig. 1. The hypocentral depths of these earthquakes, evaluated using Blake's formula (Blake, 1941), range from 10 to 15 km. The qualitative trend confirms what would have been expected on the basis of reports in the literature. In particular, Fig. 1 confirms the observation made by Howell and Schultz (1975): "inspection of isoseismal maps will suggest that the rate of attenuation of intensity with distance is often more rapid for small than for large earthquakes".

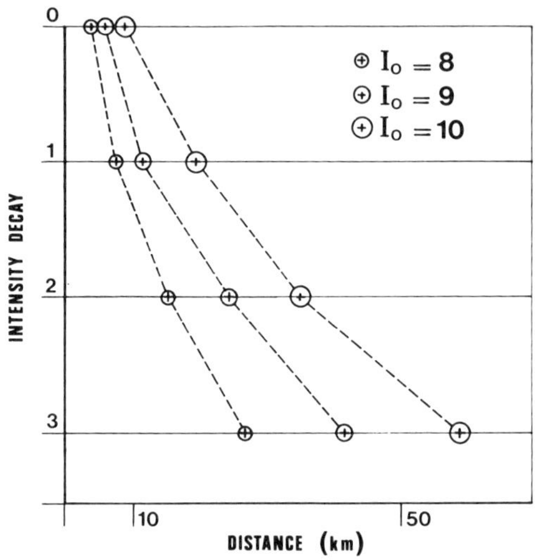

Fig. 1 *Average experimental attenuation for a sample of Italian earthquakes.*

The mean annual number λ of events and the probability distribution of epicentral intensities $F_{I0}(i) = \text{Prob}(I_0 \leqslant i)$ for the zone in which these earthquakes occurred were derived from the catalogue for the period 1680–1980. With the assumptions that the epicentres were uniformly distributed and that $F_{I0}(i)$ was constant over the zone under consideration, the seismic hazard at a particular site was calculated using two alternative attenuation laws. The experimental values of the intensity decay given in Fig. 1 were used first. Then the statistical data from the isoseismal maps were used to obtain the coefficients of the well-known attenuation law

$$I_0 - I = a + b \ln D + cD \tag{1}$$

where I_0 is the epicentral intensity and I is the intensity at a distance D from the epicentre, and the seismic hazard was calculated using this formula. The "experimental" return periods and those derived using Eqn (1) are compared in Fig. 2. The very large differences are due to the fact that according to Eqn (1) the attenuation does not depend on the epicentral intensity, whereas the experimental values show a strong dependence on I_0 (Fig. 3).

Obviously, we could use Eqn (1) with different coefficients for each intensity I_0. However, the statistical data are not generally numerous enough to ensure a good estimate of so many coefficients. It would be better to use an attenuation law with a small number of coefficients and with a structure such that the intensity decay depends on the epicentral intensity I_0 when the statistical data show this kind of behaviour.

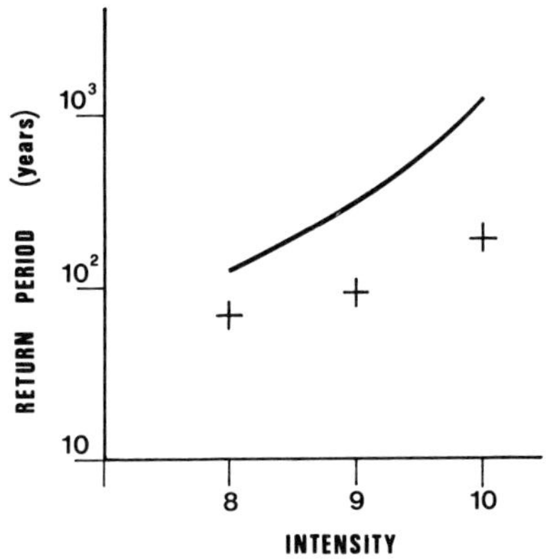

Fig. 2 *Local seismicity:* ——, *Eqn (1);* +, *experimental values.*

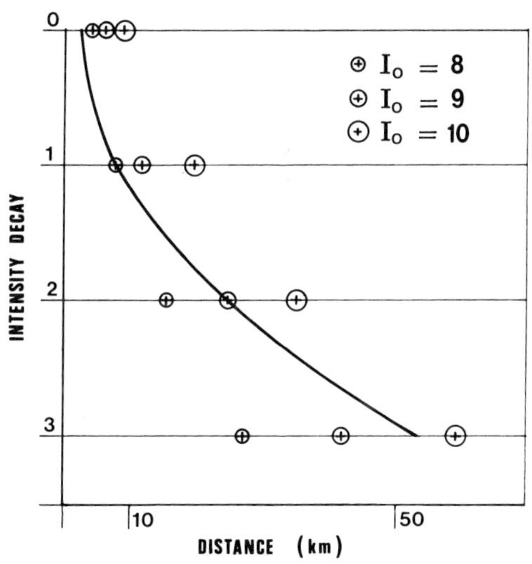

Fig. 3 *Interpretation of the sample shown in Fig. 1 using Eqn (1).*

2 A new attenuation law

The following presentation is derived from Grandori *et al.* (1987). Consider, first, the following simplified version of Eqn (1):

$$i = I_0 - I = a + b \ln D_i \tag{2}$$

In order to make the rate of attenuation depend on I_0 it is sufficient to use D_i/D_0 instead of D_i:

$$i = I_0 - I = a + b\ln\left(\frac{D_i}{D_0}\right) \tag{3}$$

where D_0 is the equivalent radius of the highest mapped isoseismal line. From Eqn (3) we obtain

$$D_i = D_0\,\alpha\exp\left(\frac{i}{b}\right) \tag{4}$$

where $\alpha = \exp(-a/b)$. Obviously, if D_0 is constant, Eqn (3) is the same as Eqn (2). However, if D_0 depends on I_0 all the radii D_i also depend on I_0. We assume that the ratio $D_0(I_0 = j)/D_0(I_0 = j-1)$ is constant (i.e. is independent of j) and denote this constant ϕ. Then

$$\frac{D_0(I_0 = j)}{D_0(I_0 = j-1)} = \frac{D_i(I_0 = j)}{D_i(I_0 = j-1)} = \phi \tag{5}$$

The constant ϕ is a measure of the rate of attenuation of small earthquakes compared with large ones. If ϕ is greater than unity, the trend is of the type shown in Fig. 1, whereas if ϕ is less than unity the rate of attenuation for large earthquakes is more rapid than that for small earthquakes. The value of ϕ for a given set of earthquakes can be calculated as the mean value of the ratios $D_i(I_0 = j)/D_i(I_0 = j-1)$. The data obtained for the sample shown in Fig. 1 give a mean value of 1.36 for ϕ.

Equation (4) implies that, for $i \geqslant 1$, the ratio

$$\psi = \frac{D_{i+1} - D_i}{D_i - D_{i-1}} \tag{6}$$

does not depend on either i or I_0. It is easy to show that

$$\psi = \exp\left(\frac{1}{b}\right) \tag{7}$$

The experimental value of ψ can be calculated as the mean value of the ratios on the right-hand side of Eqn (6). For the sample shown in Fig. 1 we obtain $\psi(i \geqslant 1) = 1.58$. For $i = 0$ the ratio ψ becomes

$$\psi_0 = \frac{D_1 - D_0}{D_0} \tag{8}$$

and taking into account the compatibility condition

$$D_i = D_0 \text{ for } i = 0 \tag{9}$$

Eqn (4) leads to

$$\psi_0 = \psi - 1 \tag{10}$$

Equation (10) is acceptable from a qualitative standpoint. In fact the rate of attenuation is faster for small distances than for large distances. However, Eqn (10) is too rigid a constraint from a quantitative point of view. For instance, the mean value of ψ_0 for the sample shown in Fig. 1 is 1.07, whereas Eqn (10) gives $\psi_0 = 0.58$. In order to overcome this inconsistency, we modify Eqn (4) by adding a new coefficient:

$$D_i = D_0\,\alpha \exp\left(\frac{i}{b}\right) + \xi \tag{11}$$

The ratio ψ is again given by Eqn (7), and the value of ψ_0 can now be adapted to the experimental data. Combining Eqns (11) and (8) with condition (9) we obtain

$$D_i = D_0\left(1 + \psi_0 \frac{\psi^i - 1}{\psi - 1}\right) \tag{12}$$

i.e.

$$\begin{aligned} i &= I_0 - I \\ &= \frac{1}{\ln \psi} \ln\left\{1 + \frac{\psi - 1}{\psi_0}\left(\frac{D_i}{D_0} - 1\right)\right\} \end{aligned} \tag{13}$$

If the coefficients ψ_0, ψ and ϕ are known, Eqns (5) and (13) completely define the attenuation law, provided that a reference value is established for D_0. This can be done as follows. Use Eqn (12) to derive a mean value $\bar{D}_0$ for each I_0 from the experimental values D_i. Then impose the condition that the reference value of D_0,

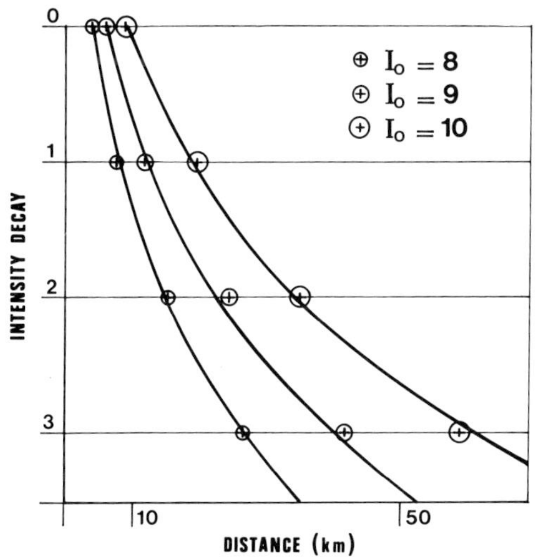

Fig. 4 *Interpretation of the sample shown in Fig. 1 using Eqn (13).*

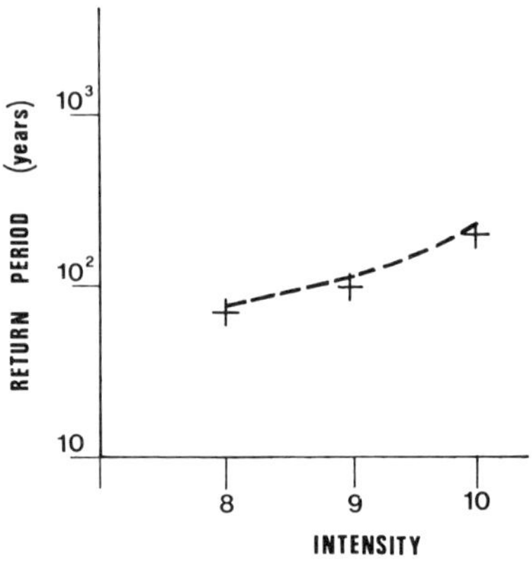

Fig. 5 *Local seismicity: ---, Eqn (13); +, experimental values.*

through Eqn (5), minimizes the squares of the deviations from the mean values $\bar{D}_0$. In this way we obtain $D_0(I_0 = 10) = 9.3$ km for the sample shown in Fig. 1.

In conclusion, the coefficients that define the attenuation law (13), with the help of Eqn (5), for the sample shown in Fig. 1 are $\psi_0 = 1.07$, $\psi = 1.58$, $\phi = 1.36$ and $D_0(I_0 = 10) = 9.3$ km. The attenuation law defined in this way fits the statistical data very well (Fig. 4). Consequently, the calculation of local hazard is also satisfactory (Fig. 5).

References

BLAKE A. (1941): On the estimation of focal depth from macroseismic data, *Bulletin of the Seismological Society of America*, **31**, 225–231.

GRANDORI G., PEROTTI F. and TAGLIANI A. (1987): On the attenuation of macroseismic intensity with epicentral distance. In CARNAK A. S. (ed.), *Ground Motion and Engineering Seismology, Proceedings of the 3rd International Conference on Soil Dynamics and Earth Engineering, Princeton, NJ, 22–24 June 1987*, Elsevier, Amsterdam.

HOWELL B. F. and SCHULTZ T. R. (1975): Attenuation of modified Mercalli intensity with distance from the epicenter, *Bulletin of the Seismological Society of America*, **65,** 651–665.

Guidelines for Earthquake-resistant Design and Construction of School Buildings in Developing Countries

TRIFUN PASKALOV

Institute of Earthquake Engineering and Engineering Seismology (IZIIS), Skopje, Yugoslavia

Abstract A code for producing seismic design regulations for school buildings in developing countries located in active seismic regions is presented. Use of the code is intended to prevent buildings from collapsing during earthquakes, protect against damage to non-structural components of the building, restrict deformability and protect against personal injury and damage to neighbouring buildings. The problems of seismic zoning for a particular country and the corresponding lateral design force are included. Legal and financial aspects of the code and training programmes for personnel employed in educational buildings are considered.

1 *Introduction*

Earthquakes are one of nature's greatest hazards to both human life and property. The loss of life in some of the world's largest earthquakes has been estimated as follows: China, 830 000 in 1556, 655 000 in 1976 and 180 000 in 1920; Japan, 143 000 in 1923; India, 300 000 in 1737, 56 000 in 1935 and 5000 in 1950; Portugal, 100 000 in 1955; Italy, 75 000 in 1908 and 1000 in 1976; Romania, 3000 in 1940 and 1700 in 1977; Morocco, 12 000 in 1960; Peru, 66 000 in 1970; Chile, 40 000 in 1939 and 25 000 in 1960; Guatemala, 23 000 in 1976; Yugoslavia (Skopje), 1073 in 1963. The damage to engineering structures and facilities caused by ground vibration, settling of foundations, landslides, land subsidence and uplift, tsunamis and fires attributed to earthquakes has been extremely large.

For example, the Balkan, Mediterranean and Middle East region, which is part of the Alpine–Himalayan seismic belt, is one of the most active regions in the world. Many of the earthquakes which occur in this area are associated with regional faults such as the East Alpine, Hellenic, Anatolian, Elbors and Zagros. Many severe earthquakes which destroyed famous cities and even civilizations in this area have been recorded in history. During the last 25 years many disastrous earthquakes have occurred, including those in Agadir, Morocco (1960), Skopje, Yugoslavia (1963), Dashte Bayaz, Turkey (1968), Banjaluka, Yugoslavia (1969), Gediz, Turkey (1970), Buinzara, Iran (1973), Friuli, Italy (1976), Bucharest, Romania (1977), Thessaloniki, Greece (1978), Kuh Banan, Iran (1978), Montenegro, Yugoslavia (1979), El Asnam, Algeria (1980), which have caused enormous loss of life and damage to property amounting to several billion dollars.

Major earthquakes are likely to occur in this area and in other parts of the world in the future. It is evident on the world map of seismic activity that the probability of occurrence of a severe earthquake in the Mediterranean and circum-Pacific seismic belts is extremely high and represents a significant risk to human life and property. Aseismic structural design plays a major role in minimizing the cost of earthquakes in terms of loss of life and damage to property.

The design and construction of facilities in seismically active zones are complex and important tasks. Site seismicity is investigated and seismic parameters are defined. The effect of these parameters on structures is also examined.

The aim of the code recommended in this paper is to produce seismic design regulations for school buildings in developing countries located in active seismic regions. The problem of seismic zoning for a particular country and the corresponding lateral force design level are included. The structural qualities of redundancy, plan symmetry, regularity and construction inspection are emphasized. Past earthquake experience has shown that the quality factor is far more important than exact evaluation of the ground motion input with respect to damage reduction. The designer should pay close attention to detail: the force transfer system should be followed carefully without omitting any stages, appropriate materials should be chosen and the workmanship should be of high quality. If these conditions are fulfilled, structures with improved seismic resistance should be obtained. Finally, local engineers, architects and planners in each country should review the proposed model code critically before accepting it. Such a review process is essential if a usable and acceptable code is to be obtained.

2 *Approach to the problem*

The purpose of an earthquake-resistant design is to provide a structure with features which will enable it to respond satisfactorily to seismic effects. These features are related to five major objectives which are listed below in order of importance.

(1) The likelihood of collapse after a very severe earthquake should be as low as possible.

(2) Damage to non-structural elements caused by moderate earthquakes should be kept within reasonable limits. Although substantial damage due to severe

earthquakes which have a low probability of occurrence is acceptable, such damage is unacceptable in the case of moderate tremors which are more likely to occur.

(3) Buildings in which many people are usually present should have deformability features which will enable occupants to remain calm even in the event of strong shocks.

(4) Personal injury should be avoided.

(5) Damage to neighbouring buildings should be avoided.

Such objectives cannot be met simply by following certain rules or recommendations. This approach can be used only in situations which occur frequently and only for their more general aspects. Each case has unique features which will require the designer to use his knowledge and experience to adapt the recommendations in such a way as to obtain the most suitable solution. Since the design of earthquake-resistant buildings involves a large number of variables, such as rigidity, damping, ductility and other structural factors whose influence on the response of the structure to earthquakes is difficult to assess, it is not possible to make specific recommendations and we can only establish guidelines for the designer.

For these reasons, before formulating specific recommendations applicable to specific types of building, it is necessary to establish general criteria which should be taken into account in order to attain each of the objectives mentioned above.

2.1 Prevention of collapse

Making a building safe from collapse appears to be simply a question of resistance. However, since earthquakes are random events and the response of buildings to them is complicated, it is obvious that the problem is much more complex and a dynamic analysis of the situation must be performed.

A building can be prevented from collapsing under static loads by ensuring that the load-bearing capacity of the structure is adequate. However, under dynamic loads, such as those caused by an earthquake, the energy which must be absorbed by the structure in order to make it fail must be taken into account. In such instances, the energy-absorption capacity of the structure is more important than its resistance. Therefore the major factor to be considered in the design of earthquake-resistant buildings is the ductility of the structure.

Although no well-defined theoretical methods for determining the ductility of a structure have yet been developed, it is useful to clarify the concept so that at least a relative appreciation of its importance can be attained and criteria for the selection of a structural scheme can be established. The greater the energy required to cause a structure to fail is, the greater is its ductility. If the resistances of two structures are equal, the more brittle structure will collapse at a lower strain, i.e. it will collapse after absorbing less energy. This means that ductile structures have a better chance than brittle structures of resisting earthquakes of unexpectedly high intensity. The ductility of a structure depends on the type of material used and on the structural characteristics of the assembly. Structures which are built using ductile materials and rigid elements are preferable to structures whose horizontal resistance is provided by non-ductile shear-resistant walls, which in turn are preferable to structures of the inverted pendulum type (such as elevated tanks) which are even less ductile.

Another factor which must be taken into account in the prevention of collapse is the hyperstaticality of the structure. Apart from those cases in which the conditions require an isostatic (statically determinate) structure to be used (for example, where there is a possibility of differential settlement), hyperstatic (statically indeterminate) structures generally have advantages which make them safer under seismic loads. If the structure is forced into the plastic yield regime under such a load, its hyperstaticality results in the formation of plastic hinges which can absorb substantial amounts of energy without depriving the structure of its stability. In other words, the plastic response of the material and the redundancy of hyperstatic structures protects such structures against unexpected stresses of unpredictable intensity such as those occurring in the case of seismic stress.

In summary, the prevention of collapse, which is the most important of the five objectives, must be sought not only by ensuring that the structure has sufficient static resistance but also by selecting materials and a structural scheme which will produce a ductile hyperstatic structure. However, this conclusion should not be interpreted as implying that the choice of building materials should be dictated solely by considerations of prevention of collapse. It is possible to obtain the desired ductility and hyperstaticality with the most commonly used building materials (steel, concrete and masonry) simply by using a suitable design. A poor design may result in a brittle structure which cannot withstand the effects of an intense earthquake.

The ductility of metal structures can be ensured by designing their joints in such a way that they will yield plastically under stresses with values above normal. Thus plastic hinges, which permit rotation, will be formed, a large amount of energy will be absorbed and the structure will not reach breaking point. Another way of achieving the same result is to avoid elastic and plastic instabilities.

Ductile structures can be built using reinforced concrete if care is taken in the design to provide the joints with sufficient abutments that can adequately confine the concrete, thus permitting it to deform plastically without breaking. It is also important to ensure that the tension edges of the structure are adequately reinforced and that there are sufficient stirrups to ensure that the concrete is properly confined along the compression edges. Moreover, it is advisable to adopt a higher safety factor in the design of reinforced concrete subjected to stresses such as shear (which cause brittle fracture) and bonding, particularly where it serves as the basis for the splices of the principal reinforcements.

Masonry can also be used to build ductile structures if an appropriate design is applied. Bricks of a suitable size and shape should be used. Relatively thin parallelepipedal bricks (i.e. bricks which will require many tiers per unit of height) are particularly suitable. The confinement of walls by reinforced concrete or steel members (confined masonry) provides the assembly with satisfactory ductility. The replacement of this confinement with vertical and horizontal reinforcements (reinforced masonry) also gives satisfactory results, particularly when the masonry is carefully assembled using construction methods that ensure perfect alignment of the reinforcement.

2.2 Protection against damage to non-structural components

Every building, in addition to its structure, has other very important elements which enable the building to function. These include partitions, windows, ducting, equipment, furniture etc., which, although they do not play a part in the building's resistance, should remain intact during an earthquake. The structural design should be such that these elements are not damaged under the action of moderate earthquakes. In order to establish what characteristics the design must have if it is to satisfy this requirement, the necessary relationships between these non-structural elements and the building's structure should be explained.

The non-structural elements generally have no stability of their own. They are usually supported by the structure and must therefore be able to move with the building without being damaged. However, the way in which the elements are connected to the structure may cause structural members to become rigid or place them under stress, and this can have serious effects in some cases, particularly when the connections are installed without detailed study or supervision. There have been many instances in which buildings were damaged because non-structural elements interfered with the behaviour of the structure itself. Thus it is necessary to have a design which takes into account the integrity of the non-structural elements when the structure is subjected to moderate earthquakes so that structural vibrations do not result in breakage of the non-structural elements and that these elements do not interfere uncontrollably with the functioning of the structure.

It is a mistake to select brittle materials for partitions or other secondary components in elastic buildings and to use them without sufficient study, as is the case with ordinary masonry in frame structures without shearing walls.

If the material has been selected with due regard to the elasticity of the structure, it will be necessary to design the separation joints and anchorages with the dual purpose of ensuring stability and at the same time avoiding any appreciable interference with the functioning of the structure. The separation joints must be of an adequate size and the location of anchorages or guides must be carefully studied. It is useful at this point to describe the solution commonly adopted when affixing masonry partitions (either structural or non-structural) in buildings which are of low or moderate height, and are therefore rigid. The solution consists in simply anchoring the masonry in the structural members of reinforced concrete which surround it (beams and columns). This solution is satisfactory if an effective anchorage can be made by using specially placed reinforcements or some other similar arrangement. It is unsuitable for elastic buildings, in which these fragile partitions should remain floating and should be supported by the structure without interfering with it. The same holds for furniture and large doors and windows. The latter should be kept floating through the use of carefully designed devices, not only to protect them from damage but also to prevent the possible consequences of their destruction.

2.3 Restriction of deformability to prevent panic

Under seismic action flexible buildings behave in a manner which has a very unfavourable psychological effect on the occupants. Rigid buildings, however, do not present this problem. The choice of whether to use a rigid or a flexible structure depends on a number of factors, one of which is the comfort of the occupants. However, except in the case of buildings which will house a large number of people, this decision is generally determined on the basis of economic and technical considerations and on the local conditions (geology of the site and the type of terrain). A flexible structure is generally preferable on rocky ground, whereas a rigid structure is more appropriate on deformable ground with a low load-bearing capacity. Greater rigidity generally results in higher cost, but this is partly compensated by the lower cost of anchorages for non-structural elements which are simpler in such cases. Rigid structures are also more comfortable for the occupants, and movable objects suffer less damage in such structures when earthquakes occur. Thus this type of structure is generally preferable for schools and other densely occupied buildings. A flexible structure is cheaper because of the lower seismic coefficients. However, this cost is increased by the higher cost of the more complicated anchoring systems required for partitions or other secondary components. Moreover, the damage caused as a result of panic and the destruction of movable property may be substantial.

As can be seen, many factors are involved and it is not easy to establish well-defined criteria. The designer, taking into account the characteristics of each case, must trade off the technical and economic factors against the comfort of the occupants.

2.4 Protection against personal injury

People inside and outside the building can be injured by falling accessories or ornamental elements (cornices, facing etc.). No earthquake-resistant design is satisfactory unless it takes this into account; although apparently secondary, it has been responsible for substantial loss of life in severe earthquakes. Again, the anchorage and attachment to structural members, both in the interior and on building façades, is important. During intense earthquakes façade coverings and the glass in large windows frequently come loose when insufficient attention has been paid to affixing them.

2.5 Protection against damage to neighbouring buildings

The main cause of damage to neighbouring buildings is collision. To ensure that this does not occur it is necessary to determine accurately how much space should be provided between buildings. This separation should be determined on the basis of the deformation which both structures will undergo and will depend on the type of structure selected. The deformation in properly designed elastic buildings with rigid frames may be three or four times that for the elastic limit of their materials. In such instances, therefore, the separation must be calculated on the basis of four times the sum of the elastic displacements plus such deformation as may result from any rotation or displacement of the foundations.

Accordingly, very tall elastic buildings should preferably be completely isolated from all other structures of similar height; if proximity is unavoidable, the distance separating them should be at least the calculated value, bearing in mind each of the possible types of displacement which might occur. It is often necessary to construct low buildings separated by short distances. In such cases, where the deformation is smaller, it is possible to make the calculated separation even smaller by using shock-absorbing components in the joints (or separation) which will reduce this type of damage to a minimum. The use of soft wood or other commercially available products has yielded good results.

These basic conditions which must be satisfied in order to meet the five objectives of earthquake engineering listed above should be taken into account in the three factors which must be considered in any construction: the architectural design, the structural design and the materials used. It should be emphasized that these factors are not independent. In modern construction, these three aspects of building design are inseparable. Studies of buildings damaged by severe earthquakes, where the damage was attributable to defects in the project, frequently reveal that the structural defects were due to poor architectural design or inappropriate choice of materials. In other cases, an excellent project designed for construction in reinforced concrete became a defective project when the structure was built in steel.

The architectural scheme will depend on the materials selected, and there will be an ideal structural scheme for that combination. Conversely, a structural form imposed by local conditions and by the materials to be used will suggest the architectural scheme most suited to the programme's requirements. It can be concluded from the foregoing that the design of a building is an interdisciplinary task in which the architect and the civil engineer must cooperate very closely in order to arrive at the most suitable architectural and structural scheme. They must constantly bear in mind the requirements of the programme, the local ecological conditions and the materials available, without losing sight of the five basic objectives of earthquake engineering.

3 Recommendations of the code

3.1 Evaluation of the seismic intensity at sites proposed for school buildings

It is important to determine the geographic coordinates, magnitudes and focal depths of earthquakes which have previously occurred within a radius of about 100 km (60 miles) of the site on which it is proposed to construct school buildings by using data available from the world seismological network, historical data and data from other sources. It is also important to determine the maximum magnitudes or intensities in the area, taking into account all available earthquake foci and using the following expression:

$$I = a + bM - c \ln\{(D^2 + d^2)^{1/2}\} \quad (1)$$

where I is the earthquake intensity in terms of the MSC-64 or MM intensity scales, M is the magnitude of the earthquake, D is the hypocentral distance (in kilometres), and a, b, c and d are constant coefficients characterizing the territory. These parameters can be improved by performing detailed seismotectonic investigations and seismic risk analysis of the territory concerned. The values of the parameters in expression (1) for the Middle East region are as follows: $a = 8.0$, $b = 1.5$, $c = 2.5$ and $d = 20.0$.

3.2 Correction of the seismic intensity according to soil conditions at the construction site (influence of soil types, underground water level and the slope of the site)

It should be noted that the seismic intensity determined as described above is assigned to soil category II. (As a model the 1981 Yugoslav Seismic Codes or the USSR SNIP-II-7-81 Codes corresponding to the following soils can be adopted: disintegrated and heavy erosive rocks; large or medium grains of gravelly sand, dense or medium dense, moist or very moist; small or very small sand grains, dense or medium dense, dry or slightly moist; clay soils.) If the soils at the construction site do not conform to the above description, the seismicity parameters should be determined as described below.

The seismic intensity at the construction site of a school building can be increased or decreased by 1 degree in accordance with the soil conditions.

(1) The seismic intensity determined using Eqn (1) should be increased by 1 degree if the construction site consists of soils of category III (loose sand with any level of moisture and grain size; saturated gravelly sand with large or medium grains; saturated or very moist sand with small or very small grains; moist plastic clay soils).

(2) The seismic intensity determined using Eqn (1) should be decreased by 1 degree if the construction site consists of soils of category I (all types of hard or moderate disintegrated rocks; moderately moist dense soils with large grains of magmatic origin, containing no more than 30% of sandy-clay infills).

In addition the seismic intensity assessed as described above should be increased by 1 degree if the school has to be built on a slope with an inclination of more than 15°.

In this case the slope should be reliable against sliding under static and seismic action. Slopes with inclinations in excess of 15° are unreliable if there are faults, heavy degraded soils due to geological processes, soils subjected to intensive settlement, rock falls, landslides, karst etc. in the vicinity. If these unfavourable conditions cannot be avoided it is important to undertake additional measures for strengthening the foundation base as well as the structure of the building.

In this code it is recommended that school buildings are not constructed on sites with seismic intensities exceeding 9 degrees. Since school buildings are very important structural systems because of the severe consequences of their destruction, the seismic intensity assumed in their design should be increased by 1 degree compared with that determined as described above.

The design seismic intensity of a particular site can be determined more accurately if detailed seismic microzoning investigations are performed.

3.3 Additional requirements of the architectural design of school buildings

The following conditions should be fulfilled in the design of school buildings in seismic regions.

(1) The construction materials selected should result in structural units with low weight but with sufficient resistance and stability.

(2) Preference should be given to symmetrical architectural and structural schemes with a uniform distribution of structural rigidities and masses due to the structural and non-structural elements of the building.

(3) When soils of categories II and III are present, the wall and column foundations should form a uniformly connected system consisting of continuous foundation bands, or the individual foundations below each column should be connected with lintel beams along the building walls.

The school building should have no more than three storeys in an area where the seismic intensity is 7 or 8 degrees, and no more than two storeys for a seismic intensity of 9 degrees.

Buildings of the corridor type are unsuitable for seismic intensities of 8 or 9 degrees.

Antiseismic joints should be used in the following cases: when the plan of the building is complicated, or when two adjacent sections of a building differ in height by 3 m or more. The antiseismic joints should be formed by constructing two independent walls, frames or other structural elements throughout the height of the building. These joints are not necessary in the foundation structure if control of differential settlement due to static loads is not required. The distance between two antiseismic joints (the length of a particular building sector) should not exceed 100 m for steel and reinforced concrete frame or shear-wall systems or 60 m for other structural systems. The minimum width of the joints should not be less than 50 mm.

The infill material in these joints should not restrict relative horizontal movements between the building sectors.

The staircase cores in the school buildings should be included in the plan of the structural system, with mandatory window openings on the front walls. The number and disposition of the staircase cores are determined by fire codes and architectural requirements, but there should not be less than one staircase between two antiseismic joints, i.e. one staircase per building section.

The construction of school buildings from adobe, mud, soil etc. in seismic regions is forbidden. The only exceptions to this rule are single-storey buildings consisting of wooden frame walls with adobe infills and a light roof structure in areas where the seismic intensity does not exceed 7 degrees.

3.4 Methodology for determining seismic forces acting on school buildings

It is assumed that in school buildings with simple geometry the seismic forces act independently in the direction of the two principal horizontal orthogonal axes. The building should be able to resist a combination of these forces and static loads. In the design of school buildings with complicated geometry the seismic forces are assumed to act in the least favourable direction for the structural system or its elements. The seismic forces in the vertical direction should be taken into consideration in the design of horizontal or inclined cantilevers and of brick or stone masonry buildings.

The seismic design force acting at a point k of a structural element can be determined by taking into account only the contribution of the first mode of vibration and applying the following expression:

$$S_k = 0.25\, k_1 k_\psi \frac{A}{g} \beta \eta_k q_k \qquad (2)$$

where S_k is the seismic force, k_1 is a coefficient taking into account the structural system of the building which takes values of 1.5 and 1.3 for frame and masonry structures respectively, k_ψ is the dissipation coefficient which equals 1.0 and 1.5 for frame and masonry structures respectively, β is a dynamic coefficient equal to 3.0, 2.7 and 2.0 for soil categories I, II and III respectively, η_k is the mode shape coefficient which can be determined from the simple formula

$$\eta_k = h_k \sum_{j=1}^{n} q_j h_j \Big/ \sum_{j=1}^{n} q_j h_j^2 \qquad (3)$$

h_k is the height of point k from the foundations of the building, q_k is the lumped mass acting at point k, A is the average peak ground acceleration, obtained on the MSK-64 intensity scale as 100 cm s^{-2}, 200 cm s^{-2} and 400 cm s^{-2} for seismic intensities of 7 degrees, 8 degrees and 9 degrees respectively and g is the acceleration due to gravity.

Cantilever elements (balconies, bay windows etc.), whose masses are negligible with respect to the total mass of the building, should be designed to carry forces determined according to Eqn (2) with $\beta\eta_k = 5.0$.

Structural elements on top of school buildings (penthouses, parapets etc.) and supports of heavy equipment should be designed to carry forces determined using Eqn (2), also with $\beta\eta_k = 5.0$.

In the design of a school building with length and/or width greater than 30 m a torsional moment about a vertical axis passing through the centre of gravity of the building should be taken into account in addition to the forces determined using Eqns (2) and (3). The eccentricity between the centres of rigidity and mass of the system should not be less than $0.02B$, where B is the building dimension normal to the direction of action of the seismic force S_k.

3.5 Additional requirements for the construction of earthquake-resistant school buildings

3.5.1 Buildings of the skeleton type

(a) The following types of skeleton systems can be used in the construction of school buildings: plane and space frame; frame with infilled walls; frame with vertical truss diaphragms; frames in combination with rigid cores. These structural systems should also be designed to withstand horizontal seismic action.

(b) The diaphragms and the rigid cores, which usually carry the largest component of the seismic force, must be constructed along the total height of the building, and should be distributed equally in both orthogonal directions and symmetrically with respect to the centre of gravity of the building.

(c) The joints of the reinforced concrete elements should be strengthened with welded reinforcement nets or closed stirrups. The parts of the beams and columns in the vicinity of the frame joints should be reinforced by closed stirrups of the required shear strength and separated by less than one cross-sectional width. This separation should be no greater than 150 mm.

(d) The staircase cores should be designed so that they can contribute to the overall stiffness of the structure or as rigid cores, capable of withstanding most of the seismic forces. Staircase cores should not be designed as separate structural units.

3.5.2 Stone or brick masonry buildings

(a) Calculations for stone or brick masonry buildings should be carried out taking into account the simultaneous action of horizontal and vertical seismic forces. The vertical seismic forces should be taken as 15%, 20% and 30% of the vertical loads for seismic intensities of 7 degrees, 8 degrees and 9 degrees respectively. The direction of action of these forces (up or down) should represent the less favourable condition for a particular structural element.

(b) The following material is suitable for carrying walls or infills in frame structures:

(1) solid or hollow brick with a compression strength not less than 7.5 MPa and holes of diameter up to 15 mm (when the seismic intensity is 7 degrees ceramic stones can be used if their compression strength is at least 7.5 MPa);

(2) solid or hollow concrete blocks, including light concrete, with bulk density at least 1200 kg m^{-3} and compression strength at least 7.5 MPa;

(3) natural blocks of limestone, with compression strength at least 3.5 MPa, or tuff (except felasite), with compression strength at least 5.0 MPa.

(c) The masonry should be erected using cement–lime mortar, lime and sand with a compression strength of at least 2.5 MPa. If blocks are used, the compression strength of the mortar should be at least 5.0 MPa.

(d) The dimensions of the elements of the stone or brick masonry wall should satisfy the calculated requirements, but they should not be less than the specifications given in Table 1.

Table 1

Seismic intensity (deg)	7	8	9
Minimum width of columns (m)	0.70	1.00	1.35
Maximum width of wall openings (m)	3.50	3.00	2.50
Ratio of walls to openings	0.33	0.50	0.75
Dimensions of cantilevers in walls (m)			
Same material	0.20	0.20	0.20
Reinforced concrete	0.40	0.40	0.40
Wooden elements	0.75	0.75	0.75

(e) Reinforced concrete lintel beams should be erected in each wall at floor or roof level. These beams should be strongly connected at each joint. The beam width must be the same as the width of the wall, except for the lintel beams on the façade walls which could be 50 mm shorter to allow for thermal insulation. The thicknesses of all these beams must be at least equal to the thickness of the floor structure, and in any case no less than 200 mm.

(f) Vertical reinforcement elements (columns) must be included in all masonry school buildings constructed in regions with seismic intensities of 8 and 9 degrees. These elements must be erected at each corner of the building as well as at the free end of each wall.

(g) The minimum reinforcement for the horizontal lintel beams is four bars of diameter 10 mm for a seismic intensity of 7 or 8 degrees, and four bars of diameter 12 mm for a seismic intensity of 9 degrees. The minimum reinforcement for the vertical columns is four bars of diameter 12 mm for a seismic intensity of 7 or 8 degrees, and four bars of diameter 14 mm for a seismic intensity of 9 degrees.

4 Local and administrative aspects of the formulation of the code

4.1 Legal aspects

If the technical measures outlined above are to be put into effect, certain legal, administrative, fiscal and financial measures will need to be taken in earthquake-prone countries. These may include fiscal and financial incentives to invest in the protection of educational buildings and to improve contingency planning for possible disasters.

National legislation should provide for the organization of protective and emergency measures in two stages: (1) pre-disaster planning, i.e. development plans at local and regional level, which should define measures for the protection of educational buildings against possible earthquakes; (2) emergency measures, including technical measures that should be taken in the event of an earthquake.

Existing legislation may prove inadequate to deal with the situation created by a large-scale disaster. A corpus of specific legislation must therefore be prepared in advance, which will take effect after a disaster as soon as the government declares a state of emergency in the affected area. All relevant legislation should be made available in a single document to facilitate quick reference and rapid action in emergencies. In particular, such legislation should provide for the following:

(1) an immediate freeze on all real-estate transactions;
(2) temporary limitation of private property rights;
(3) review and eventual revision of the land tenure system;
(4) allocation of responsibilities for emergency action.

4.2 Financial and fiscal measures

It is useful to distinguish between financial provisions for long-term protection measures and those for repair and strengthening after a disastrous earthquake. In the former case, provision can be made in national legislation for tax incentives, low interest loans and, where possible, graded insurance premiums. Such measures would provide the government with means to exert pressure on the owners of educational facilities to take appropriate long-term protective action. In the latter case, financial assistance will be needed to meet the cost of repair and strengthening. In some cases, this may be forthcoming through collective efforts of the national or international community.

5 Training and information programmes for personnel employed in educational buildings

Many recent earthquake catastrophes have revealed that the general population is not prepared for disasters. There is an evident need to develop appropriate policies of education and public information in disaster-prone areas. Appropriate measures to be taken include the following.

(1) Education on the nature and effects of violent natural phenomena, such as earthquakes, and on methods of protection against them should be introduced into physics and geography syllabuses in schools.

(2) Special training courses designed to prepare occupants of school buildings for the effects of earthquakes should be included in the curriculum. The aim of these courses should be to prevent panic in the event of a disastrous earthquake and to enable the occupants to take appropriate action to protect the buildings from secondary consequences such as fire, flood etc.

(3) Appropriate education on aspects of earthquake engineering and engineering seismology relevant to the protection of educational facilities in earthquake-prone areas should bc provided at vocational and university level (e.g. in university courses in architecture and civil engineering).

(4) Studies of physical planning in disaster-prone areas should be introduced into university curricula.

(5) Workshops and training courses should be organized at national or regional level for specialists in the protection of important buildings in earthquake-prone areas.

There is at present a marked lack of information on the behaviour of various types of school buildings in the event of disasters, and further research is needed in order to understand and predict how particular buildings will react to a strong earthquake. In particular, research on structural and soil dynamics should be directed specifically to the problem of stabilizing the structures and foundations of existing educational buildings.

6 Application of the recommendations in the code

Since the code is designed for developing countries, the implementation of its recommendations is not a simple task. There are a number of reasons for this, among which the following are the most relevant.

(1) Each developing country must be treated as a particular case. The implementation of any code of practice should be performed in the context of the priorities of the country in question.

(2) If the recommendations given in Section 3 are to be effective in a particular developing country, they must be adapted to the specific conditions existing in that country. The level of exposure to seismic hazards, the local building traditions and materials, and the qualified manpower available must be taken into account.

(3) The effectiveness of application of a code of this type depends strongly on the education and qualifications of the engineers in the design offices, particularly those responsible for controlling the quality of construction. It is very important that the inspectors of construction on site are adequately remunerated.

(4) An adequate code of practice for any developing country, based on the recommendations given here, could be drawn up by a specialized research organization or by a group of experts qualified in seismology, earthquake engineering and engineering seismology.

(5) Finally, we recommend that UNESCO should provide all possible assistance in promoting the adoption of these recommendations in developing countries through education and public information. In particular, UNESCO could assist by setting up appropriate educational programmes at the vocational and university level and by organizing workshops in close collaboration with inter-governmental and non-governmental organizations such as the International Association of Earthquake Engineering.

Bibliography

AOYAMA H. (1981): Outline of earthquake provision in the recently revised Japanese Building Codes, *Bulletin of the New Zealand National Society for Earthquake Engineering*, **14** (2), 63–80.

APPLIED TECHNOLOGY COUNCIL (1978): Tentative provisions for development of seismic regulations of buildings, *ATC Publication 3–06*.

GOVERNMENTAL COMMITTEE OF USSR FOR CONSTRUCTION WORKS (1982): Design codes of structures in seismic regions, *SNiP II-7-81*, Part II, Chapter 7, Sarita Prakashan, Meerut, India.

HOUSNER G. W. (1977): Response of structures to ground shaking: Theme Report on Topic 3, *6th World Conference on Earthquake Engineering, New Delhi, January 10–14, 1977.*

INTERNATIONAL ASSOCIATION OF EARTHQUAKE ENGINEERING (1981): *Basic Concepts of Seismic Codes*, Lisbon.

INTERNATIONAL ASSOCIATION FOR EARTHQUAKE ENGINEERING (1980): *Basic Concepts of Seismic Codes*, Vol. 1, Tokyo.

INTERNATIONAL ASSOCIATION FOR EARTHQUAKE ENGINEERING (1982): *Basic Concepts of Seismic Codes*, Vol. 2, Tokyo.

INTERNATIONAL ASSOCIATION FOR EARTHQUAKE ENGINEERING (1984): *Earthquake Resistant Regulations—A World List*, Tokyo.

KARNIK V. and ALGERMISSEN S. T. (1978): Seismic zoning. In *Assessment and Mitigation of Earthquake Risk*, pp. 11–47, UNESCO, Paris.

KRISHNA J. (1978): Concepts for evaluation of seismic design codes for buildings and civil engineering structures, *Publication No. 61*, IZIIS, Skopje.

KUBO K. and KATAYAMA T. (1978): Earthquake resistant properties and design of public utilities. In *Assessment and Mitigation of Earthquake Risk*, pp. 171–184, UNESCO, Paris.

PASKALOV T. and ZELENOVIC V. (1984): *Code for Design and Analysis of Civil Engineering Structures in Seismic Regions*, Yugoslav Federal Bureau of Standards, Belgrade.

PETROVSKI J. (1978): Seismic microzoning and related problems. In *Assessment and Mitigation of Earthquake Risk*, pp. 48–65, UNESCO, Paris.

PETROVSKI J. (1978): Earthquake engineering achievements and development of unified seismic design codes, *1st Arab Seismological Seminar, Baghdad, December 1978*, University of Baghdad, Baghdad.

STANDARDS ASSOCIATION OF NEW ZEALAND (1976): *Code of Practice for General Structural Design and Design Loadings for Buildings*, Standards Association of New Zealand, Wellington.

YUGOSLAV FEDERAL BUREAU OF STANDARDS (1980): *Code of Practice for Earthquake Resistant Design and Construction of Building Structures*, Yugoslav Federal Bureau of Standards, Belgrade.

ZSUTI T. C. and SHAH H. C. (1978): *Recommended Seismic Design Provisions for Guatemala*, John A. Blume Earthquake Engineering Center, Stanford University, Stanford, CA.

Methods of Increasing the Reliability and Economic Efficiency of Construction in Seismic Regions

E. A. SEKHNIASHVILI AND M. A. MARJANISHVILI

Institute of Computer Mathematics, Academy of Sciences of the Georgian SSR, Tbilisi, USSR

Abstract Methods of reducing the damage caused by earthquakes are discussed. There are two fundamental approaches to reduction of the seismic load: decreasing the mass of the building without reducing its strength characteristics, and decreasing the dynamic coefficient. These methods are discussed with the aid of examples, and some recommendations are made.

A number of strong earthquakes occur every year in which tens of thousands of people perish and material damage totalling several hundred million dollars occurs. The construction of public and residential buildings (apartment houses, schools, hospitals, theatres, office buildings etc.) which provide maximum safety (reliability) at a reasonable cost is a major and complex problem because these requirements are conflicting and consequently mutually exclusive. However, this problem can be successfully solved in earthquake engineering because, if all other conditions remain unchanged, the external seismic load on the structure depends on the properties of the building, its structural characteristics and the materials from which it is constructed.

In the USSR the magnitude of the seismic load S_{ik} at a point k in a given direction and corresponding to the ith harmonic of the natural oscillation of a building is calculated using the formula

$$S_{ik} = k_1 k_2 k_\psi A Q_k \beta_i \eta_{ik} \tag{1}$$

where Q_k is the mass of the building acting at point k, A is a coefficient determined by the seismicity of the region where the building is located, k_1, k_2 and k_ψ are empirical coefficients that take into account the acceptable damage to a building and its structural characteristics, β_i is a dynamic coefficient depending on the period of natural vibrations of the building, its dissipative properties and the soil conditions, and η_{ik} is a coefficient depending on the mode of deformation of the building during free vibration. It should be noted that the formulae adopted to determine the value of a seismic load are in principle analogous to Eqn (1) in many countries, although this equation obviously does not include all possible seismic impacts on buildings and their reaction to these impacts.

Nevertheless, Eqn (1) suggests two fundamental approaches to protection from earthquakes, i.e. to reduction of the seismic load. The first approach is to decrease the mass Q_k of the building without reducing its strength characteristics, and the second is to decrease the dynamic coefficient. We shall now consider the first approach.

Assume that two identical buildings are constructed. The first is a standard structure with a bearing framework (e.g. columns in a braced frame or inner bearing walls in a panel building without frames) constructed from "traditional" material (steel, ordinary reinforced concrete). The second building, in which the bearing framework is made of low modulus material (e.g. constructional aluminium, constructional plastic, light-weight reinforced concrete etc.), has strength characteristics analogous to those of the "traditional material" but with a modulus of elasticity which is a factor of n less than the modulus of elasticity of the "traditional material".

The dynamic coefficient in Eqn (1) is given by

$$\beta_i = \frac{N}{T_i} \tag{2}$$

where $1 \leqslant N \leqslant 1.5$ (depending on the soil) and $0.8 \leqslant \beta_i \leqslant 3$. The period of flexural-and-shear free vibrations of a structure of height H, considered as a distributed-parameter system, is given by the formula

$$T_i = \alpha_i \left(\frac{mH^4}{EI} \right)^{1/2} \tag{3}$$

where m is a characteristic of the inherent mass, α_i is a coefficient and EI is the flexural rigidity of the structure. If the modulus of elasticity E of a low modulus material decreases more rapidly than its three-dimensional mass m, T_i will increase and β_i and S_{ik} will decrease. Consequently, when a light-weight low modulus material is used, the seismic load S_{ik} will decrease not only in direct proportion to the reduction of the three-dimensional mass but also as a result of the increase in the flexibility of the structure as a whole. This formula also shows that E should be reduced more rapidly than m, and not vice versa.

Figure 1 shows the vertical section and the floor plan of an 11-storey building constructed from ordinary and light-weight reinforced concrete for a region with a seismicity of 7 degrees, and Table 1 shows the design seismic loads for this building (the asterisks indicate the loads for the light-weight reinforced concrete). It is obvious from the table that the ratio of the masses $\Sigma Q^*_k/\Sigma Q_k$ is 0.775, the ratio of the

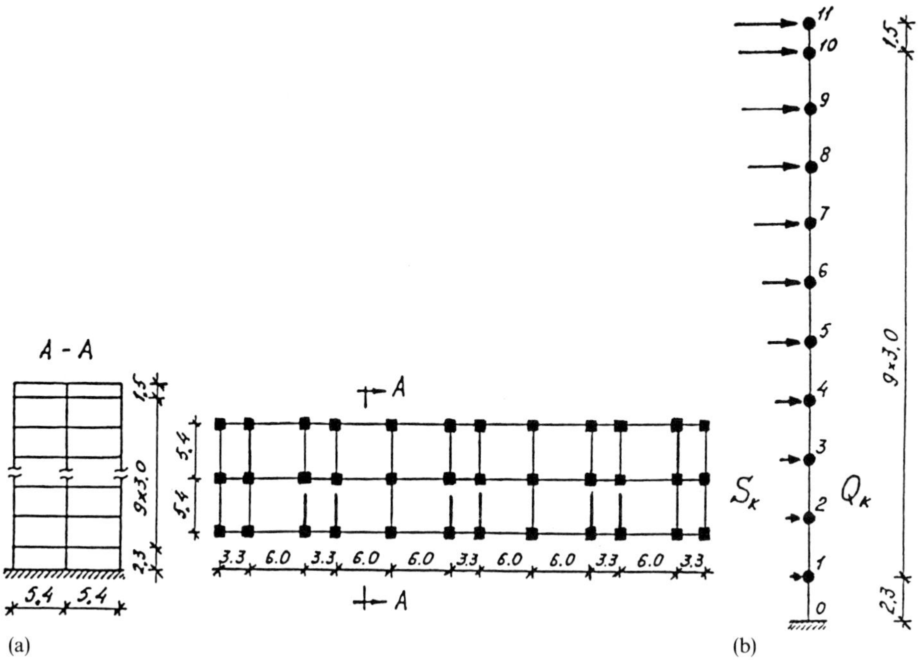

Fig. 1 *Vertical section and floor plan of an 11-storey building constructed from ordinary and light-weight reinforced concrete.*

Table 1 *Design seismic loads for the 11-storey building shown in Fig. 1*

Floor	Q_k (kN)	Q_k* (kN)	S_k (kN)	S_k* (kN)	M_k (kN m)	M_k* (kN m)
11	6429	4550	368.6	201.5	—	—
10	5013	3620	270.5	142	553	302
9	6530	5142	305.7	187	2470	1333
8	6530	5142	258	158.3	5305	2924
7	6530	5142	210	129.2	8913	4991
6	6530	5142	163	100.5	13 151	7445
5	6530	5142	119	73.2	17 879	10 200
4	6530	5142	79	48.4	22 963	13 175
3	6530	5142	45	27.3	28 285	16 296
2	6530	5142	19.3	11.3	33 741	19 498
1	6804	5300	4.1	2.1	39 255	22 734
0	—	—	—	—	43 492	25 219
	ΣQ_k = 70 486	ΣQ_k* = 54 606	ΣS_k = 1842	ΣS_k* = 1081		

Q_k and S_k are defined in Fig. 1.

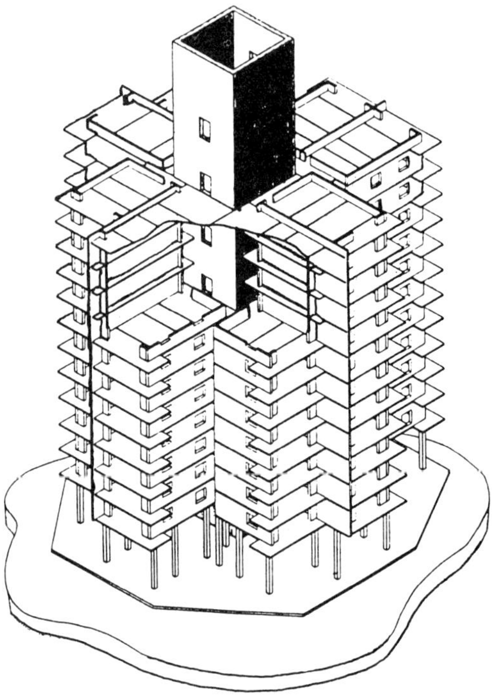

Fig. 2 *Building with a central core of monolithic reinforced concrete surrounded by panel constructions.*

seismic loads $\Sigma S^*_k/\Sigma S_k$ is 0.587 and the ratio M^*_k/M_k of the foundation bending moments caused by the seismic loads is 0.580. Therefore the use of low modulus materials, particularly light-weight concrete and light-weight reinforced concrete, provides a direct method of active seismic protection. Furthermore, porous aggregate or material that can be used in its manufacture are often widely available in seismic regions.

It is clear that in this case the bearing framework of the building will work under more favourable conditions not only because of the reduction in the vertical load but also because of the decrease in the eccentricities of the application of vertical (normal) forces so that edge stresses in the components of the framework of the braced frame will decrease. Consequently, the cross-sections of the components of the bearing framework can be reduced and less material can be used, or the cross-sections can be kept the same size and materials with lower strength characteristics can be used. These procedures enable the cost of the building to be reduced. This approach is effective for building monolithic and prefabricated structures.

The second approach is to seek engineering solutions for structures which will enable buildings (a) to restructure and adapt to an earthquake, (b) to have variable parameters and, if necessary, dampers and absorbers of vibrational energy, and (c) to show high dissipative properties etc. Prefabricated and unified components and products are used in the mass construction of frame and frameless panel buildings in the USSR. The development of industrialized processes for construction using uniform components has increased the speed of building, improved quality and minimized seasonal fluctuations. In prefabricated buildings it is comparatively easy, using various design techniques, (a) to reduce the values of seismic forces acting on a building, (b) to adapt buildings, (c) to localize plastic deformation in individual components of a building, (d) to use elastic and friction joints between the

Fig. 3 *16-storey tower block.*

components and (e) to produce an artificial increase in the dissipation of vibrational energy. This approach has been used in recent years in the USSR, particularly the Georgian SSR, to develop, investigate, design and build new types of large panel structures for mass housing.

A building consisting of a central core of monolithic reinforced concrete surrounded by panel constructions is shown in Fig. 2 to illustrate one of the engineering solutions. Such structural solutions allow improved architectural expression for the elevations of large panel buildings, and the panels operate as a damper during earthquakes. A 16-storey tower block with a cross-shaped plan is shown in Fig. 3. The main support of the building is a monolithic reinforced concrete core (made using slip forms) without battens or embedded fittings. Floors, which can be regarded as plane discs, fit tightly round the core although they are not connected to it.

Figure 4 shows an example of a 14-storey apartment block with prefabricated monolithic cores. This building has a complex configuration consisting of two displaced half-sections, with the cores at the centre. The cores are constructed from prefabricated reinforced concrete panels 0.3 m thick connected at the corners by monolithic wells (Fig. 5). Plane floor slabs 0.16 m thick, of room size, are rigidly

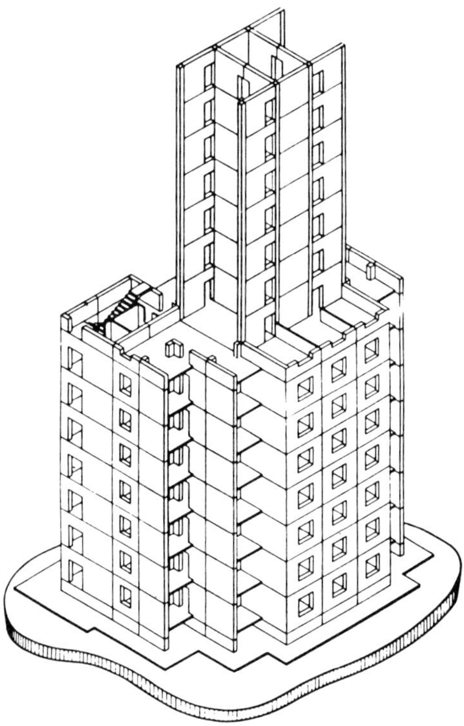

Fig. 4 *14-storey apartment block with prefabricated monolithic cores.*

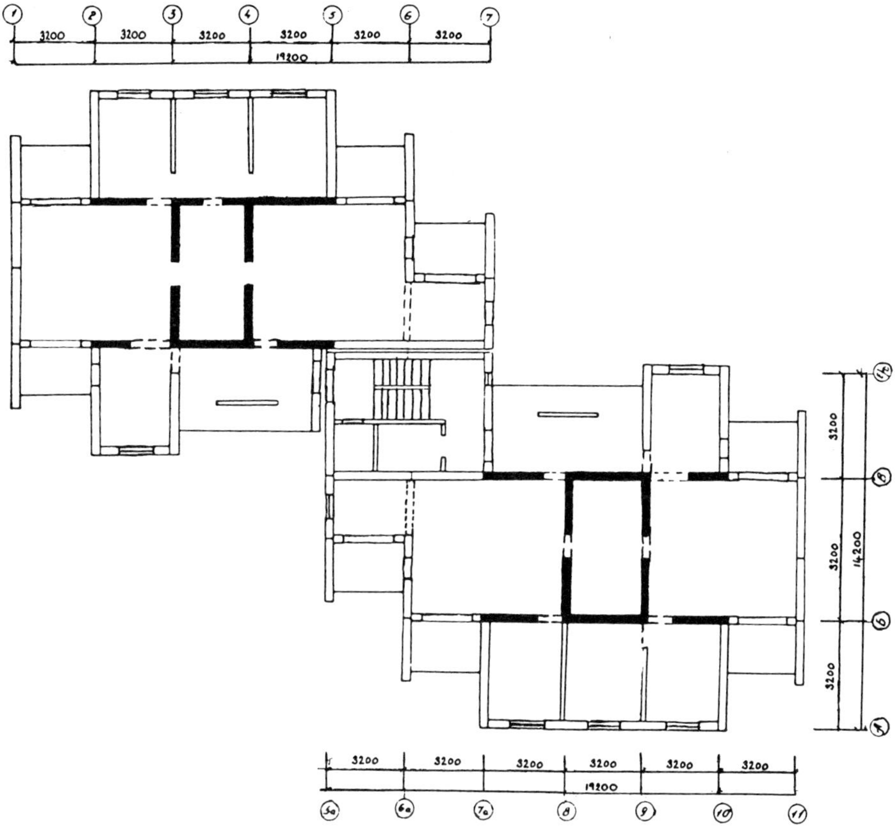

Fig. 5 Cross-section of the apartment block shown in Fig. 4.

connected to the core and interconnected with each other, thus forming a rigid horizontal disc. The outer wall panels are made of ceramsite concrete, and the connections between wall panels and between wall panels and floors are flexible. Dynamic tests performed on a building of this type using a powerful vibration machine which caused first and second resonance oscillations showed fairly good agreement between the experimental and design characteristics of the building.

For extensive large panel buildings it is advisable to use a structural solution incorporating damping storeys. The building consists of several cores connected every three to four storeys by rigid floors. The other floors are flexibly connected to each other and to the cores. The wall panels also have flexible ties, allowing displacements which show the effect of dry friction. Flexible storeys can damp the vibrations produced by seismic impacts (Fig. 6). Buildings of this type provide active seismic protection. At particular values of the horizontal seismic load they can restructure and change their dynamic parameters, and this results in a reduction of the negative effects of earthquakes.

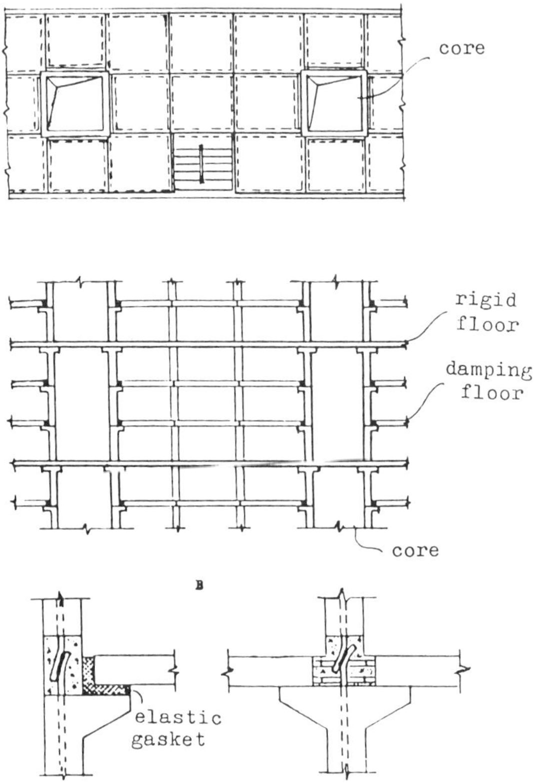

Fig. 6 *The use of flexible storeys to damp vibrations.*

Cable-and-trunk structures, which have been widely used in recent years, also provide active seismic protection (Fig. 7). Such buildings consist of a core and a suspended main part (interstorey floors etc.) which can be considered as a multimass system connected to the core by elastic ties. Under seismic impact the core vibrates and the whole of the suspended part is drawn into the vibrational process. If the floors are rigidly joined to the core they vibrate synchronously with the suspended part, but if the floors are not connected to the core then only the core vibrates and the suspended part is protected from the action of seismic loads and the consequent displacements. However, if the displacement of the core is large it may collide with the floor supports. In addition, buildings of this type are sensitive to wind loads. Therefore in the most acceptable version of such a building the interstorey floors and the core are connected by an elastic tie, resulting in a system with dynamic dampers.

If the cables from which the main part of the building is suspended are prestressed and fastened to the foundation, the rigidity of the core and hence of the whole building will increase. Consequently, the dynamic parameters of a building

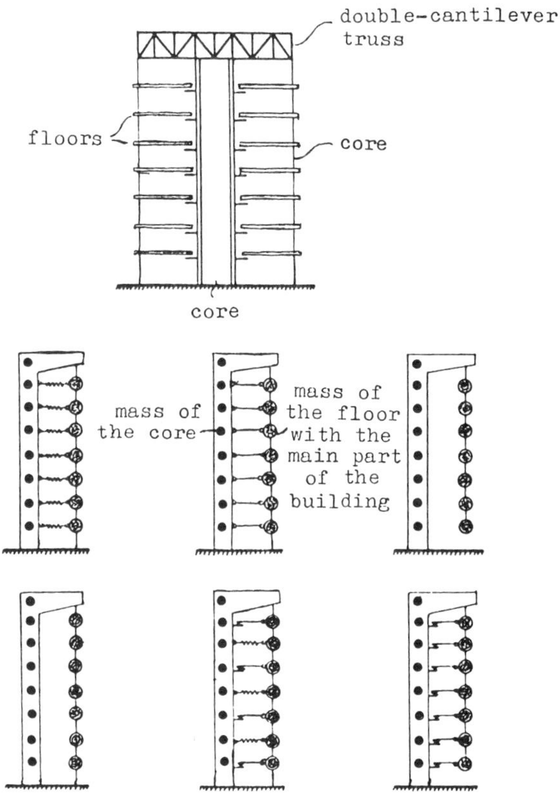

Fig. 7 *Cable-and-trunk structure.*

can be adjusted by prestressing the cables and a dynamic damper tuned to the expected seismic impact can be obtained. If the prestressed cables break at the foundation level under a given seismic load, the rigidity of the whole building will change immediately and the seismic load will decrease because the structural system of the building has changed owing to the effect of the disconnected ties. As a result, the main part of the building will be freely suspended and it will not take part in the core vibrations. The cables can be prestressed again easily and speedily.

Experimental tests of the building types described here were carried out using two models. One of them consisted of a core with a head in the form of a horizontal steel cantilever frame to which suspensions (wire ropes) were fastened; the main part of the building (floors etc.) was attached to these ropes. The second model consisted of a core surrounded by reinforced concrete columns on which the main part of the building (floors etc.) rested. The core had a cross-section of dimensions 75 cm × 75 cm, a height of 400 cm and a wall thickness of 5 cm. At the level of the interstorey floors the core walls contained embedded fittings to support the floor slabs.

Static and dynamic load tests were carried out on models with a vertical load on the floors. Dynamic tests were conducted using a vibration machine placed above the core. These tests were carried out in stages. The dynamic characteristics of the core were determined first, and after the suspension of the main part of the building (floors etc.) the complete system was tested. The models were tested in two ways: (a) without fastening the suspensions to the bedplate, and (b) with fastening and prestressing. Various methods of attaching the floor slabs to the core were investigated: hinged support of slabs to the core; rigid attachment of slabs to the core; alternating rigid attachment and free support; elastic gaskets in the gaps between the floor slabs and the core.

The experiment showed that when the floor was freely supported the core did not take part in vibrations in the resonance regime, and when the floors were rigidly attached to the core they vibrated together with the core in the resonance regime. The displacement of the core was 3.5–4 times higher when the floors were rigidly fastened than when they were freely supported.

As a result of these investigations it is recommended that in practice rigid attachment of floors to the core should be alternated with free support. The smallest displacements of floors and cores in the resonance regime of vibrations were obtained with this arrangement.

Energy Absorption Elements for Application in the Construction of Earthquake-resistant Buildings

DIMITAR JURUKOVSKI, BORIS SIMEONOV AND VICTOR TRAJKOVSKI

Institute of Earthquake Engineering and Engineering Seismology (IZIIS), Skopje, Yugoslavia

Abstract A seismic-energy-absorbing steel element which can be used in the construction of multistorey steel frame buildings in seismic or windy areas has been developed. Quasi-static tests of this element were carried out under laboratory conditions using a full-scale single-storey model. Actual gravity loads acting in the columns of the basic frame system and cyclic horizontal loads representing the horizontal seismic effect were simulated. The effectiveness of elements of this type was studied by using them in a hypothetical building design.

1 *Introduction*

The behaviour of high-rise structures during earthquakes depends on their strength capacity and ductility factor. However, the accumulation of deformations increases the degree of damage of the structures, which in turn increases the level of risk to humans and materials. Earthquake damage to structures can be decreased by dissipating the released energy, and the use of energy-absorbing elements for this purpose has been studied. These elements can either be built into some of the structural components or be used as special elements forming an integral part of the structure. The use of energy-absorbing elements enables two goals to be achieved. First, the increase in the energy-absorption capacity and the structural rigidity results in a decrease in the level of deformation and the degree of structural and non-structural damage in the building. Second, the quantity of built-in material is reduced and hence the mass of the building is decreased, resulting in a more rational construction.

Quasi-static tests have been carried out on a full-scale steel frame constructed from reinforced concrete floor structures with built-in steel energy-absorbing elements made of hollow structural sections (HSSs). The main aims of these tests were as follows: (a) to determine the non-linear behaviour of the frames with energy-absorbing elements and to examine the level of dissipated energy, the ultimate load-carrying capacity and the failure mechanism; (b) to obtain the data required to develop an adequate mathematical model and a procedure for predicting the non-linear behaviour, which would be a step towards the final application of this type of energy-absorbing element in real structures in seismic areas.

2 *Test programme*

A five-storey building was designed with steel energy-absorbing elements built in one span of two frames over the whole height of the building in both orthogonal directions. The prefabricated frame columns were made of steel HSSs of dimensions 260 mm × 260 mm × 7 mm filled with concrete, and the floor structures were reinforced concrete slabs 20 cm thick. One frame containing energy-absorbing elements was selected from this structure for quasi-static testing. Dynamic analysis of the building design was performed according to the Yugoslav Code for Aseismic Design and Construction, and the loading schedule for the tested frames was determined on the basis of this analysis.

The effectiveness of the element in increasing the energy-absorption capacity and rigidity was evaluated by performing quasi-static tests on the same frame without energy-absorbing elements. The simulated gravity loads on the columns and the conditions for supporting the tested element and for load simulation were identical for the two cases.

2.1 Design of test models

The test model has a floor height H of 3.0 m and a span L of 3.5 m (Fig. 1). The columns are steel HSSs of dimensions 260 mm × 260 mm × 7 mm filled with concrete. I14 profile beams are bolt-connected to the columns and embedded in the floor structure, which is 20 cm thick, between a grid reinforcement 8/10 cm in diameter in the lower zone and a grid reinforcement 12/11 cm in diameter in the upper zone. The compression strength of the concrete used in the floor structure is 30 MPa and the width of the slab is 90 cm.

The energy-absorbing elements are made of HSSs of dimensions 120 mm × 120 mm × 5 mm. The energy-absorbing elements in the internal portion of the frame span are rectangular with dimensions of 108 cm × 95 cm and are connected to four cross-sections, each 107 cm long, directed diagonally outwards. All the joints in the energy-absorbing elements are welded. All profiles and energy-absorbing elements are made of steel with a yield point of 240 MPa.

The geometry of the frame model without energy-absorbing elements is shown in Fig. 2. Two models with this geometry were constructed and tested. Figures 1 and 2 show that the frame geometries in the two cases considered are identical.

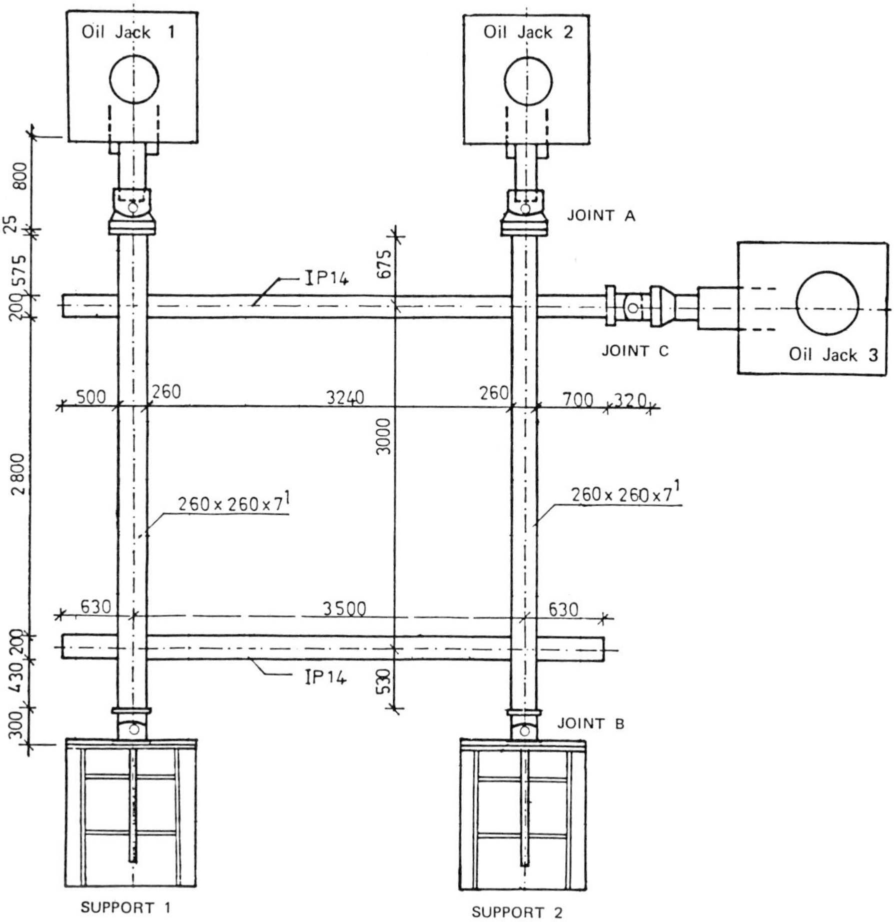

Fig. 1 *Experimental set-up of the frame model with an energy-absorbing element (dimensions in millimetres).*

2.2 Inelastic behaviour of the frames

The inelastic behaviour of the frames can best be explained using hysteresis diagrams of the lateral load versus the displacement at the upper slab level of the model (Figs 3 and 4). The development of cracks in the upper and lower slabs when the tensile stress is exceeded is the first sign of inelastic behaviour.

The characteristic yield point of the model with energy-absorbing elements was determined on the basis of the hysteretic relationship between the lateral load and the displacement at the point when the rigidity of the model underwent a marked decrease. During this state the strains in the tensile reinforcement of the slab exceed those of yielding, and neither local nor lateral buckling occurs in the steel columns and the energy-absorbing elements. After yielding, the lateral force continues to

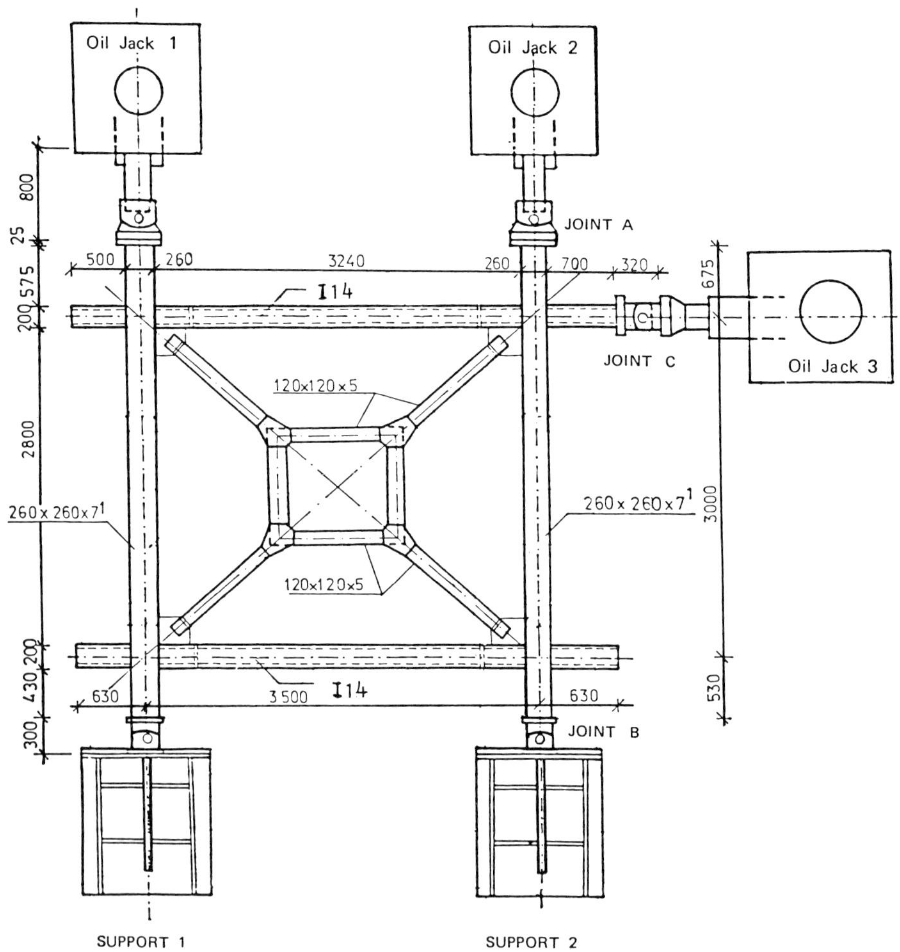

Fig. 2 *Frame model without an energy-absorbing element (dimensions in millimetres).*

increase and reaches its maximum value at a lateral displacement of 20 mm when local buckling occurs at the ends of the horizontal parts of the rectangle in the centre of the steel absorbing elements. After a few cycles in this state the lateral force decreases and cracks develop in the steel profiles of the energy-absorbing elements at the points where local buckling had previously been observed.

In the final phase of the test the lateral displacement is increased to 32 mm with a substantial decrease in the lateral force and the rigidity of the model. The cracks in the steel energy-absorbing elements increase but rupture does not occur.

The force–displacement relationship for the frame model without energy-absorbing elements is shown in Fig. 5. It can be seen that at $\delta = 0.25$ cm the elastic force reaches a value of 20 kN and the model starts to behave non-elastically. The

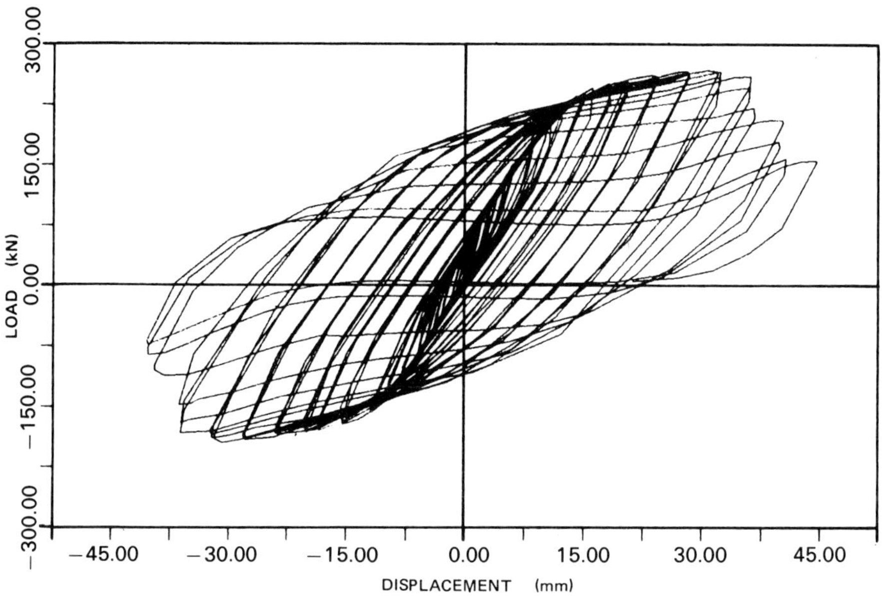

Fig. 3 *Lateral load–displacement hysteresis loops (model CR20M1).*

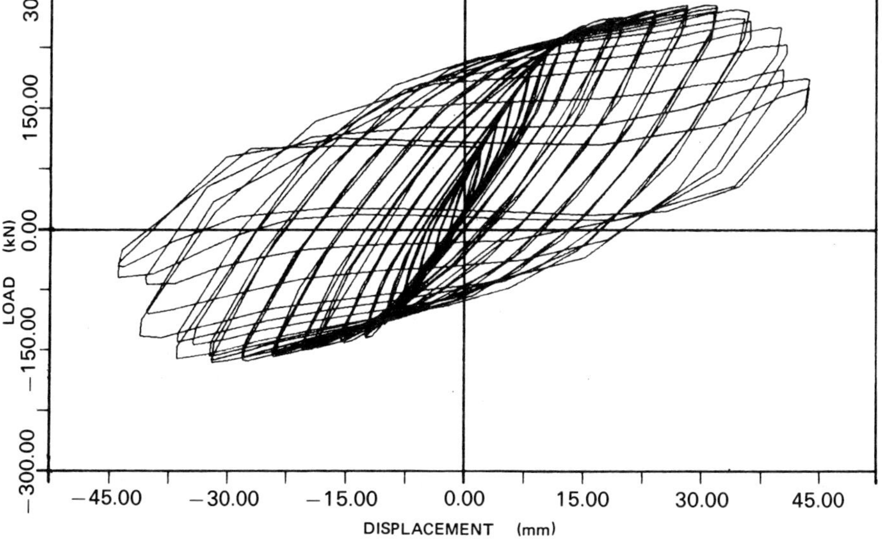

Fig. 4 *Lateral load–displacement hysteresis loops (model CR20M2).*

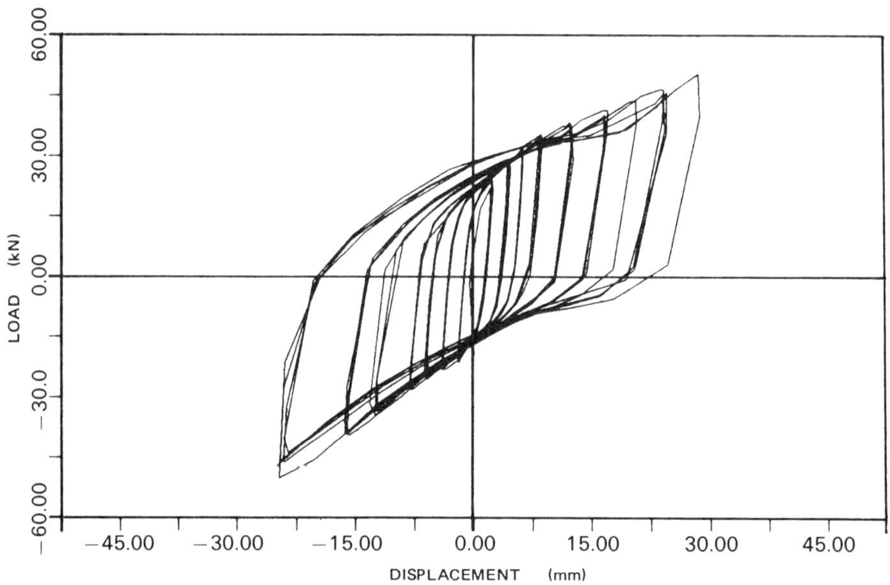

Fig. 5 *Lateral load–displacement hysteresis loops (model CR10M1).*

first cracks appear in the floor slabs and they increase further as the force increases. Over the range of deformation tested ($\delta = 3.0$ cm) the columns remain in the elastic regime.

2.3 Frame strength

An earthquake-resistant structure should be able to maintain its load-carrying capacity for cyclic horizontal loads in the inelastic regime. Also, after a strong earthquake, a structure should still have sufficient strength capacity to sustain the gravity loads and lateral forces of the aftershocks. The experimental results obtained from tests of the two full-scale frames with energy-absorbing elements meet the above requirements.

A constant inclination of the hysteresis loops, implying constant rigidity, is observed prior to yielding. The yield strength is estimated as the mean value of the lateral force in the two loading directions and is found to be 190 kN. Because of an initial inaccuracy in the frame geometry different lateral loads are observed in the positive and negative loading directions. After yielding, the lateral force continues to increase, reaching a mean maximum value of 225 kN at a lateral displacement of 20 mm. When the lateral displacement was increased to 28 mm the strength capacity of the models decreased to 170 kN, which is 75% of the maximum attainable strength capacity.

In the final phase of model testing the lateral force decreases on each cycle, indicating the absence of stable hysteresis loops. This is due to the development of cracks in the steel profiles in the central part of the energy-absorbing element.

If we compare the force values given in Figs 3 and 5 for the same value of deformation we find that, at a deformation of 3.0 mm, the strength of the frame with energy-absorbing elements is almost twice that of the frame without such elements. Also, the maximum strength capacity is approximately six times greater. This shows that the energy-absorbing elements not only substantially increase the absorption capacity of the frame but also have a marked effect on changes in the strength capacity. In addition they influence the behaviour of the frame under a cyclic force.

2.4 Displacement ductility of the frames

The deformability of a structure can be expressed in terms of the ductility factor (the relationship between the maximum horizontal displacement and the yield displacement). However, this concept has a disadvantage: it is impossible to determine the yield displacement precisely (Mahin and Bertero, 1976). The yield displacement can be estimated from the initial inelastic strain somewhere in the structure either at the onset of lateral and local buckling in an element or at the onset of a marked decrease in lateral rigidity.

It has been mentioned previously that the yield displacement of the models with energy-absorbing elements was determined from the hysteresis diagrams at the point where a marked decrease occurs in the lateral rigidity of the model. The displacement ductility for the yield displacement determined in this way is $D\delta = 3.30$. The lateral displacement for this ductility is 28 mm when the lateral force decreases to 75% of the maximum force applied during the testing. This lateral displacement state is completely acceptable in view of the strength capacity of the models at maximum displacement.

We can therefore conclude from the above results that, under the given cyclic displacement history, the models remain stable for gravitational and lateral loads at a maximum lateral displacement of 28 mm.

2.5 Dissipation of frame energy

These tests can also be used to investigate the capacity of the models to absorb and dissipate energy. In both cases the dissipated energy is determined from the hysteresis diagrams of lateral force versus lateral displacement (Figs 3 and 4). To enable comparison of the dissipated energy, it is normalized with respect to the mean lateral displacement in the same cycle. Figure 6 shows the normalized dissipated energy for both models in the non-linear regime. It can be seen that most of the energy is dissipated at lateral displacements between 20 and 30 mm. The maximum energy dissipated per cycle is 420 kN, which is comparable with the energy dissipated in a three-storey model with reinforced concrete walls with a total height of 290 cm and cross-sectional dimensions of 7 cm × 153 cm (Simeonov, 1982).

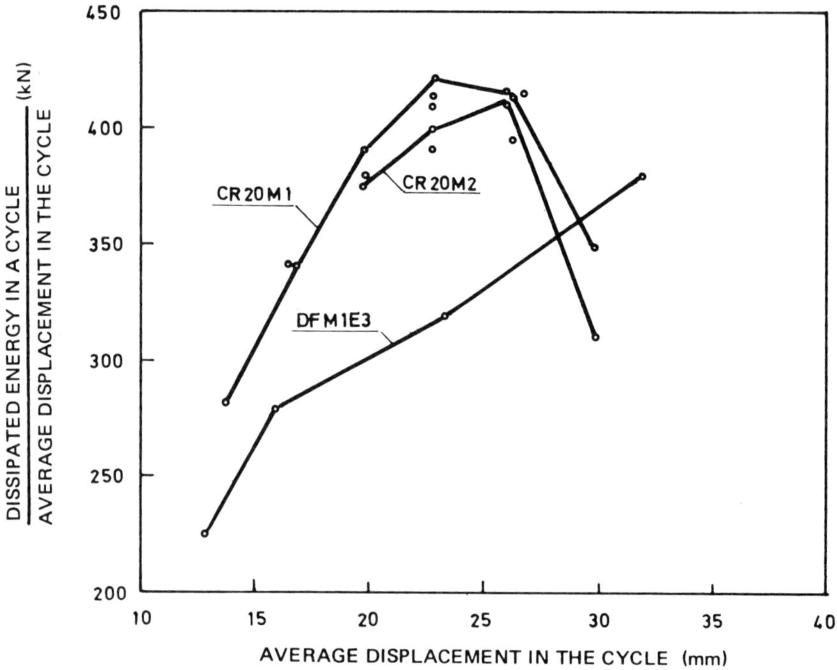

Fig. 6 *Normalized dissipated energy.*

3 *Analysis of a hypothetical building*

A five-storey building with the characteristic plan shown in Fig. 7 was chosen for practical analysis. The seismic-energy-absorbing elements are located in the shaded zones. The building is constructed of hollow steel square profiles of dimensions 260 mm × 260 mm × 6.3 mm filled with concrete, separated by distances of 360 cm and interconnected by IP12 steel profiles that are subsequently embedded in a reinforced concrete slab of thickness 20 cm. Diagonal steel dampers are located at four levels in the first and fourth spans of frame R1Y and in the second and fourth spans of frame RX3 and are connected at the centre by a square of cross-section 100 mm × 100 mm and thickness 4.0 mm. The partition walls are constructed of light prefabricated panels.

The building was analysed for the following four earthquake movements, each normalized to acceleration levels of 0.12*g* and 0.15*g*:

Montenegro 1979 earthquake, record taken at Petrovac, NS component;
Montenegro 1979 earthquake, record taken at Ulcinj, NS component;
Montenegro 1979 earthquake, record taken at Ulcinj, NS component;
Montenegro 1979 earthquake, record taken at Bar, NS component.

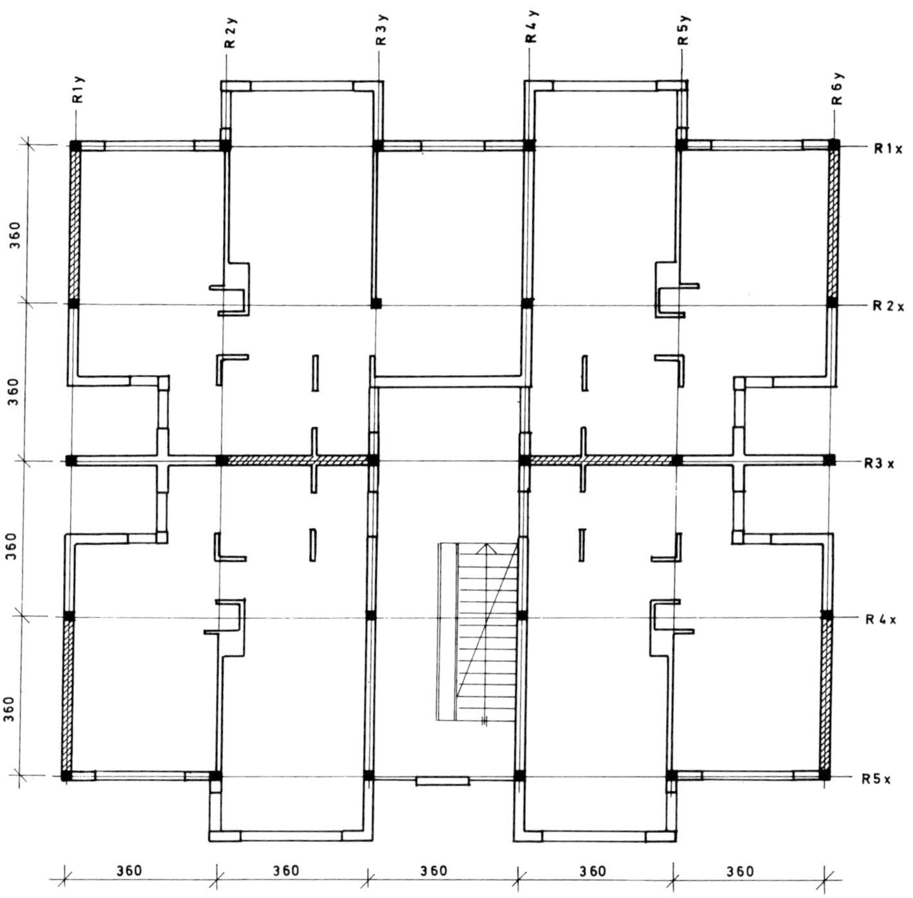

Fig. 7 *Characteristic plan (dimensions in centimetres).*

The results obtained from the analysis for the two orthogonal directions, expressed in terms of ductility, are shown in Figs 8 and 9 respectively.

4 Conclusions

The results obtained from tests of two full-scale frames with steel energy-absorbing elements can be summarized as follows.

(1) When models containing steel HSSs as energy-absorbing elements ($D/t = 24$) are subjected to cyclic loads they can develop substantial inelastic deformation and dissipate significant quantities of energy. During the inelastic load

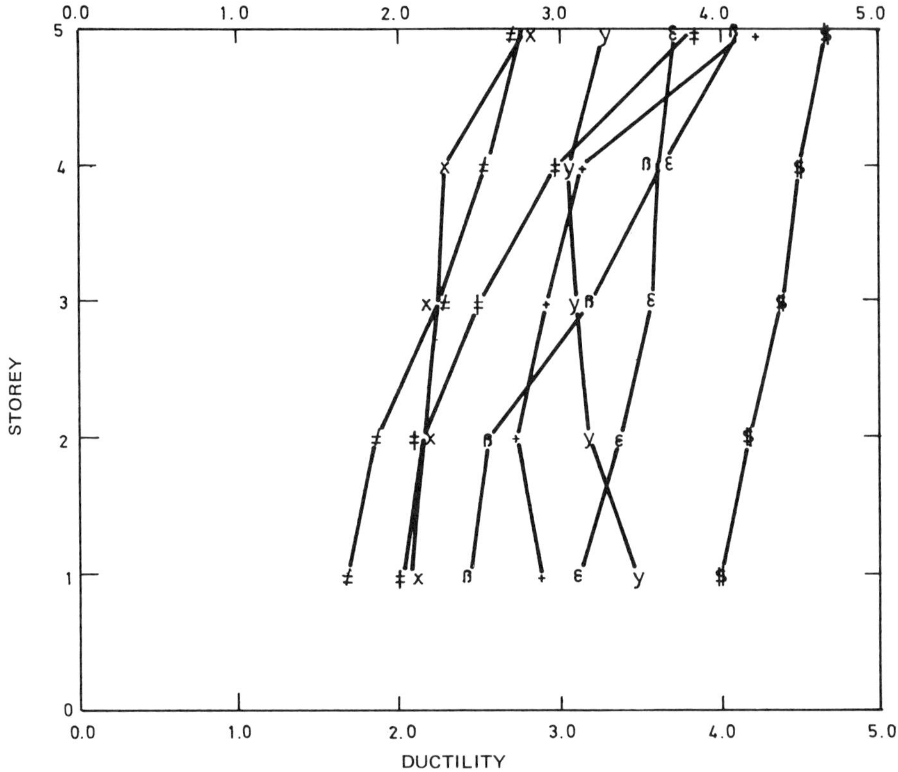

Fig. 8 *Required ductility in the XX direction: ‡, Petrovac (Oliva), NS, 0.12*g*; +, Petrovac (Oliva), NS, 0.15*g*; ×, Ulcinj (Olimpik), NS, 0.12*g*; γ, Ulcinj (Olimpik), NS, 0.15*g*; ≠, Ulcinj (Albatros), NS, 0.12*g*; β, Ulcinj (Albatros), NS, 0.15*g*; ε, Bar (Sobranie na Opstima), NS, 0.12*g*; $, Bar (Sobranie na Opstima), NS, 0.15*g.

cycles the models dissipate more than 110 times the energy that would have been absorbed in the yield state.

(2) Lateral buckling of the energy-absorbing elements is not observed during the test. This is very important with respect to their application in residential and public buildings in seismic regions.

(3) After the maximum horizontal force has been reached, local buckling occurs in the central part of the energy-absorbing elements. This causes a slight decrease in the strength capacity, but the energy dissipation capacity increases so that the hysteresis loops are stabilized up to a horizontal displacement of 28 mm.

(4) The proposed system of combined steel and reinforced concrete elements for residential and administrative buildings in seismically active regions enables bearing structures as well as partition walls and façade walls to be constructed rapidly and economically using a large number of precast elements.

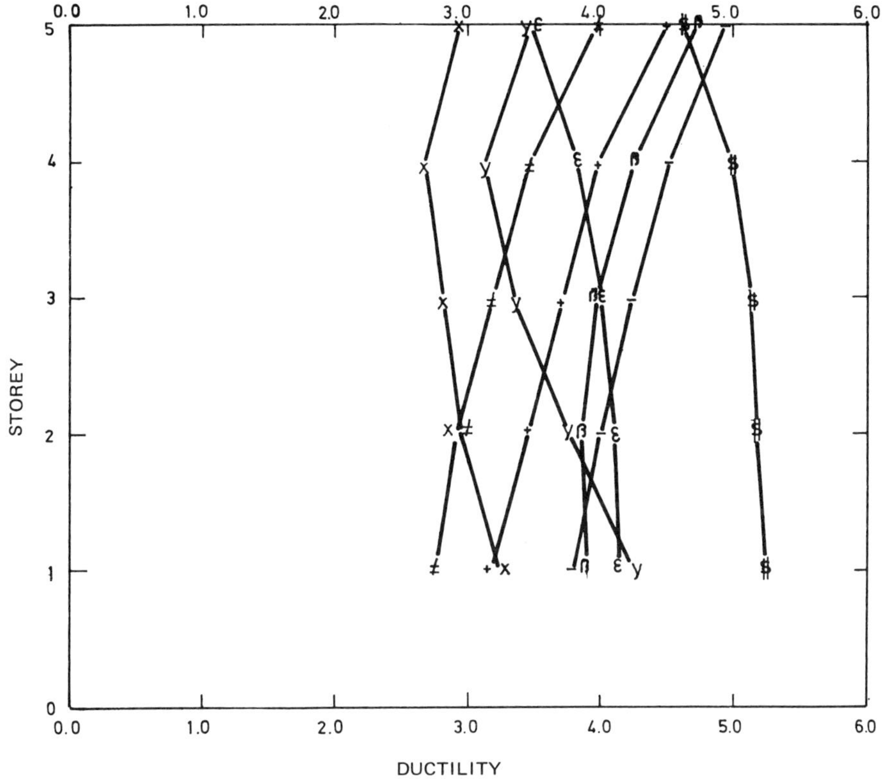

Fig. 9 *Required ductility in the YY direction (symbols as in Fig. 8).*

References

MAHIN S. A. and BERTERO V. V. (1976): Problem in establishing or predicting ductility in aseismic design, *International Symposium on Earthquake Structure Engineering, University of Missouri-Rolla, St Louis.*

SIMEONOV B. (1982): Linear and nonlinear behaviour of reinforced concrete structural walls in multi-storey buildings, *Doctoral Thesis*, Civil Engineering Faculty, University of Belgrade.

Bibliography

TYLER R. G. (1978): A tenacious base isolation system using round steel bars, *Bulletin of the New Zealand National Society for Earthquake Engineering*, **11** (4), 273–281.

TYLER R. G. (1978): Tapered steel cantilever energy absorbers, *Bulletin of the New Zealand National Society for Earthquake Engineering*, **11** (4), 282–294.

TYLER R. G. (1983): Preliminary tests on an energy absorbing element for braced structures under earthquake loading, *Bulletin of the New Zealand National Society for Earthquake Engineering*, **16** (3), 201–212.

TYLER R. G. (1985): Further notes on a steel energy-absorbing element for braced frameworks, *Bulletin of the New Zealand National Society for Earthquake Engineering*, **18** (3), 270–279.

ZAYAS V. A., POPOV E. P. and MAHIN S. A. (1980): Cyclic inelastic buckling of tubular steel braces, *Report No. UCB/EERC-80/16*, Earthquake Engineering Center, University of California, Berkeley, CA.

Collision Diagrams for a Single Degree of Freedom

M. DIKMEN

Boğaziçi University, Istanbul, Turkey

Abstract A phase diagram is used to illustrate the collision of a non-linearly elastic damped system with a rigid obstacle for a single degree of freedom.

1 *Introduction*

An operator formulation of the elastic collision of elastic systems with a rigid obstacle has been given by Dikmen (1987a). In that formulation neither a driving force nor the effect of damping were taken into account. Specifically, a discretized model consisting of springs and a finite number of mass points colliding elastically was considered. In the present paper a single degree of freedom is considered, for simplicity, to illustrate the behaviour of a non-linearly elastic system with damping and energy loss during collision. Numerical integration of the equations is used for this purpose, and the phase plane representation proves to be a very simple device which provides an adequate description.

The simple model of elementary mechanics for colliding particles (material points) has been and still is used satisfactorily in various engineering and physics problems. One such class of problems concerns the study of ergodicity (Dikmen, 1987b). The mathematics involved has varying degrees of complexity. In the case of collisions induced by earthquakes, the problem can be simplified because only the first few impacts need to be considered, in contrast with other vibro-impact problems in which the investigation must be extended over a longer period of time.

2 *The differential equation*

The mechanical behaviour of a non-linearly elastic system with viscous damping and one degree of freedom is represented by the well-known second-order differential equation

$$m\ddot{x} + b\dot{x} + k(x) = f(t) \tag{1}$$

where x is the coordinate corresponding to the degree of freedom, m is the mass, b is a damping factor, $k(x)$ represents the spring force and $f(t)$ represents the disturbing force. As usual $\dot{x}$ and $\ddot{x}$ represent first- and second-order differentiation with respect to time t. Other types of energy dissipation, such as dry friction or turbulent damping, could also be included in Eqn (1) but are not considered here.

In those cases in which $k(x)$ may be expanded in a power series, a satisfactory approximation may be obtained by retaining only the first few terms, thus reducing $k(x)$ to a polynomial which can be restricted to third-order terms, i.e.

$$a_0 + a_1 x + a_2 x^2 + a_3 x^3$$

and the origin of the variable x can be chosen so that the fixed term vanishes. If, furthermore, the spring force is assumed to be symmetric with respect to this origin, $k(x)$ takes the form

$$K(x + Nx^2)$$

where K is the spring constant of the corresponding linearized system and N is a constant for the non-linear part of the spring force. If we now introduce the natural frequency

$$\omega_0 \equiv \left(\frac{K}{m}\right)^{1/2}$$

of the corresponding linearized system, a dimensionless time

$$\tau \equiv \omega_0 t$$

can be defined. Furthermore, the right-hand side of Eqn (1) can be rewritten in the form

$$mE\,F(t)$$

where E is a representative acceleration corresponding to the disturbing force. In earthquake engineering problems, E can conveniently be chosen to be the maximum acceleration induced by the motion of the soil.

The dimensionless form of Eqn (1) is

$$\frac{\mathrm{d}^2 X}{\mathrm{d}\tau^2} + \beta\frac{\mathrm{d}X}{\mathrm{d}\tau} + X(1 + \varepsilon X^2) = \gamma\Phi(t) \tag{2}$$

where the dimensionless variable X is given by

$$X \equiv \frac{\omega_0{}^2}{g} x$$

the dimensionless damping coefficient β is given by

$$\beta \equiv \frac{b}{m\omega_0}$$

the coefficient ε of non-linear elasticity is given by

$$\varepsilon \equiv N\frac{g^2}{m{\omega_0}^4}$$

and the dimensionless disturbing force is given by

$$\Phi(\tau) \equiv F\{t(\tau)\}$$

The ratio of the earthquake acceleration E to the gravitational acceleration g is denoted

$$\gamma \equiv \frac{E}{g}$$

Obviously, $\beta > 0$ for an energy-dissipating system, and $\varepsilon > 0$ or $\varepsilon < 0$ for systems with stiffness increasing or decreasing respectively with X.

For the purposes of numerical computation and representation in the phase plane, Eqn (2) can be rewritten in the form of a system of two first-order equations:

$$\begin{aligned} \frac{\mathrm{d}X}{\mathrm{d}\tau} &= Y \\ \frac{\mathrm{d}Y}{\mathrm{d}\tau} &= -X-\varepsilon X^3-\beta Y+\gamma\Phi(\tau) \end{aligned} \tag{3}$$

The initial conditions are

$$\begin{aligned} X(0) &= 0 \\ Y(0) &= 0 \end{aligned} \tag{4}$$

and

$$\Phi(0) = 0 \tag{5}$$

Y is the dimensionless speed, which must be multiplied by g/ω_0 to obtain the physical speed. For numerical computation ω_0, ε and β are determined experimentally as characteristics of each particular system, while $f(t)$ and hence $\Phi(\tau)$ are taken from actual accelerogram records.

However, for a better insight it is useful to consider $f(t)$ or $\Phi(\tau)$ functions periodic in t or τ. The simplest of these, which satisfies the requirement that it vanishes for $t = 0$, is $\sin(\omega t)$ or

$$\Phi(\tau) = \sin(\eta\tau)$$

where

$$\eta \equiv \omega/\omega_0$$

i.e. the ratio of the excitation frequency to the natural frequency of the system. Then Eqn (2) becomes

$$\frac{d^2X}{d\tau^2}+\beta\frac{dX}{d\tau}+X(1+\varepsilon X^2)=\gamma\sin(\eta\tau) \tag{6}$$

This is the well-known Duffing equation (with damping effect). This equation was investigated first by Duffing (1918) himself and subsequently by many other workers. Despite its apparent simplicity, its solution, even in the numerical form, is far from trivial. Extensive information on classical solutions for the Duffing equation is given by Davis (1962) and Kauderer (1958). Equation (6) can also be expressed in the form

$$\begin{aligned}\frac{dX}{d\tau}&=Y\\ \frac{dY}{d\tau}&=-X-\varepsilon X^3-\beta Y+\gamma\sin(\eta\tau)\end{aligned} \tag{7}$$

together with the initial conditions (4) and (5).

3 *The collision conditions*

The collision conditions consist of a constraint imposed on the coordinate X, e.g. the unilateral constraint

$$X \leqslant H \tag{8}$$

(where H is a constant), which is adopted here, and the discontinuity condition

$$Y^+ = -eY^- \qquad \text{at } X = H \tag{9}$$

where Y^- and Y^+ are the values of Y immediately before and after the collision.

The coefficient of restitution e obeys the inequalities

$$0 \leqslant e \leqslant 1 \tag{10}$$

where $e = 1$ corresponds to elastic collision and lower values of e correspond to energy-dissipating impacts. It is well known that e is not really a constant and that it depends on the speed immediately before the collision. The method described in Section 4 allows us to use a variable coefficient of restitution:

$$e = e\left(\frac{g}{\omega_0}Y^-\right) \tag{11}$$

4 *Phase diagrams*

A differential equation like Eqn (1) can satisfactorily describe the behaviour of a

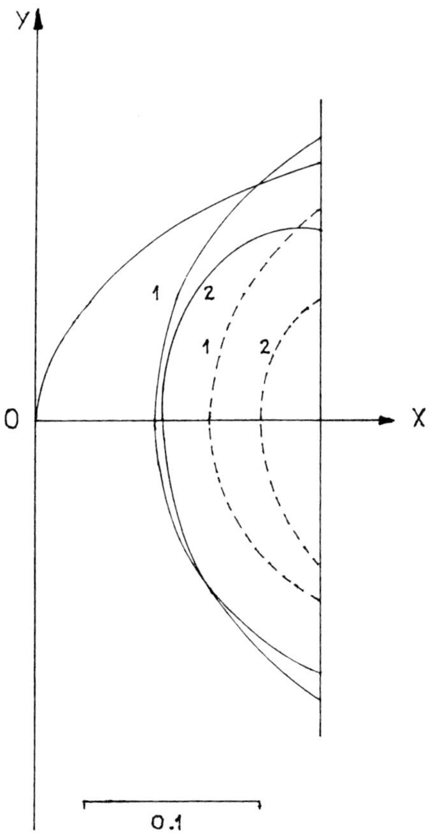

Fig. 1

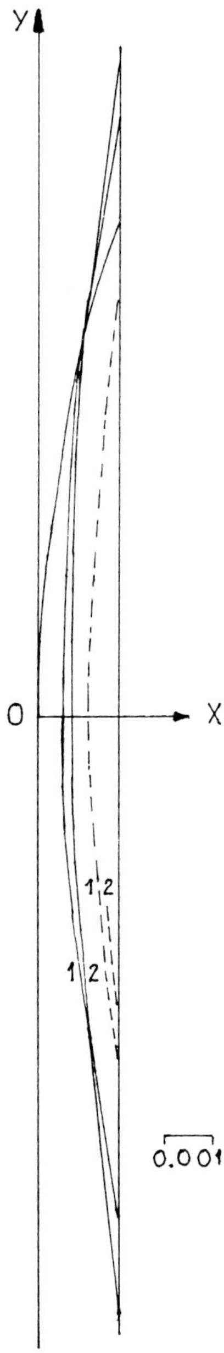

Fig. 2

system as long as the elasticity and damping characteristics of the system remain unchanged. If destructive collisions occur a short time after a structure is set in motion, then it will not make sense to investigate the nature of the solutions after a short time interval, as has already been pointed out in Section 1.

The phase trajectory starting from the origin in the phase plane at the beginning of the motion will meet the line $X = H$ at a point with ordinate $Y = Y_1$. Then a discontinuity will arise and the phase trajectory will restart (at the same time) at the point $(H, -eY_1)$. The phase trajectory must be followed until it again meets $X = H$, at ordinate Y_2 say. Again there is a discontinuity in the phase trajectory and a restart at $(H, -eY_2)$. This behaviour continues with restarts at $(H, -eY_n)$ etc. At each of these discontinuities e can be given values depending on $Y_1, Y_2, \ldots, Y_n$ etc., as in Eqn (11). At each collision the kinetic energy of the system is equal to $\frac{1}{2}(gY_n/\omega_0)^2$, i.e. proportional to Y_n^2.

In a broader study of the Duffing equation several numerical experiments have been performed for various system characteristics and excitation functions. As a simple illustration, Fig. 1 shows diagrams obtained for Eqn (7) with $\beta = 0.05$, $\varepsilon = -0.04$, $\gamma = 0.25$ and $\eta = 0.35$; H was given the rather large value of 0.162. The initial conditions were specified as in Eqns (4). The full curves after the first collision correspond to $e = 1$ and the broken curves correspond to $e = 5/7$, which are both kept fixed. Figure 2 shows the diagrams for the same values of the parameters but with $H = 0.001\,62$.

5 Concluding remarks

The phase diagram representation has been introduced above for the case of a single degree of freedom. It is obvious that the corresponding phase space can be used for representation for more degrees of freedom. However, the advantage of a simple visualization is then lost. Bilateral constraints can also be taken into account.

References

Davis H. T. (1962): *Introduction to Nonlinear Differential and Integral Equations*, Dover Publications, New York.

Dikmen M. (1987a): Some mathematical remarks on the problem of elastic collision. In Koridze A. (ed.), *Engineering Aspects of Earthquake Phenomena*, pp. 91–96, Omega Scientific, Wallingford.

Dikmen M. (1987b): Some remarks on a problem of ergodicity, *Bulletin of the Technical University of Istanbul*, **40**, 141–152.

Duffing G. (1918): *Erzwungene Schwingungen bei veränderlicher Eigenfrequenz und ihre technische Bedeutung*, Vieweg, Braunschweig.

Kauderer H. (1958): *Nichtlineare Mechanik*, Springer, Berlin.

Engineering Definition of Seismic Action Using Seismograms

A. H. BARBAT, J. A. CANAS AND J. M. CANET

Civil Engineering School, Polytechnic University of Catalonia, Barcelona, Spain

Abstract The procedures generally used to simulate seismic motion in zones where actual records are not available employ models based on the theory of stochastic processes to generate artificial accelerograms. In this study, however, seismic action is defined through deterministic and stochastic response spectra which are obtained by using the information provided by seismograms. By using spectral analysis, it has been possible to obtain a formula which relates the peak values of the pseudo-acceleration spectra to magnitude, anelastic attenuation and epicentral distance. The maximum ground acceleration thus obtained and the mean predominant frequency of the earthquake motion are then used to develop deterministic response spectra. The amplitude Fourier spectrum and the derived power spectral density obtained from the Fourier amplitude spectrum of the observed seismograms are used to define the seismic action by means of stationary and non-stationary stochastic seismic response spectra. A procedure which permits the generation of families of artificial accelerograms that match the proposed response spectra is also described.

Examples are included to illustrate the procedures described. Finally, the proposed methodology is applied to the seismic zone of the Pyrenees, which is a typical seismic zone without acceleration data.

1 *Introduction*

The design computation of structures subjected to seismic ground motions requires the structural engineer to make decisions on the definition of the action to be considered in the analysis. This operation is conditioned by the quantity and quality of seismological data available for a given area and, at the same time, by the way in which the seismic response of the structure to be analysed is expressed. This response can be characterized as a time history through its maximum value in the complex frequency domain or by means of a stochastic formulation. In the type of definition used for the response it must be feasible to take into account the possible non-linear structural behaviour under seismic action. Consequently, it is necessary to use analysis procedures which can solve both linear and non-linear structural models.

From the point of view of the definition of seismic action an "ideal" seismic zone is one for which reliable models of the characteristics of the ground motion at the epicentre can be developed and for which high quality data on the expected amplitude, frequency and duration of the ground motion are available. The existing data must also include the mechanical characteristics of the soil layers at a given site. If strong ground motion accelerograms are also available for such a zone, these can be used directly in the computation of the seismic response of structures. Numerical stepwise integration procedures for the equations of motion can be performed to compute the time history of the structural response in the linear or the non-linear field. The same acceleration records could be used to calculate seismic response spectra, in which case the maximum response of structures is computed using a modal analysis.

If ground acceleration records are not available for the seismic zone, artificial accelerograms can be generated. These are accelerograms simulated using mathematical models based on the theory of the stochastic process and can be generated by starting from the data mentioned above. Seismic response spectra corresponding to the simulated accelerograms can also be computed. Then the structural analysis procedures discussed above can be applied to this case.

For seismic areas with low seismicity and poor or no instrumental acceleration data, the numerical definition of the seismic action to be used in earthquake engineering computations has to be performed in a completely different way. Thus approximate response spectra can be estimated in a deterministic analysis by starting from some general data like the maximum ground acceleration and the predominant frequency of the expected earthquake.

In a stochastic analysis, procedures can be developed which permit the extreme mean value of the response of a single-degree-of-freedom model to be computed by solving the first-passage problem. The maximum expected response is similar to a linear seismic response spectrum ordinate, which permits the definition of **stochastic response spectra**. In this case the seismic ground acceleration can be modelled as a white noise, a filtered white noise or a non-stationary random process which is obtained by filtering a white noise process through a deterministic intensity–time function. The input data required for this type of model are the power spectral density of the acceleration of the desired ground motion and the characteristics of the filters necessary to adapt the generated spectra to some given local ground conditions.

In both deterministic and non-deterministic cases of definition of the response spectra, the maximum response of structures is computed by applying a modal uncoupling of the equations of motion. However, it must be emphasized that numerical procedures allowing the generation of families of independent accelerograms which match the prescribed spectra are available. This fact enables the time history of the seismic response of structures to be computed even in this case, in which the seismic response spectrum concept is used to define the seismic action.

In this study the definition of seismic action in zones in which the available instrumental data are basically seismograms is considered. Generally speaking, seismograms are time histories of the displacement, velocity or acceleration of the ground motion. In most cases the word "seismogram" is used to describe displacement or velocity, and accelerogram to describe acceleration. In the following the term "seismogram" will be used only for displacement or velocity records.

In regions of moderate seismicity or with poor or no accelerograph distribution it is quite impossible to obtain acceleration data. Therefore the only approach to solving the problem is to work with seismograms. A procedure which permits the maximum acceleration and the predominant frequency to be obtained using seismograms is described. Problems involving seismic wave attenuation due to anelastic effects, which are directly related to the different types of terrain, are also discussed in the text.

The shape of the Fourier spectrum is as important as the accuracy in the determination of the position of its peaks. If the shape of the spectrum is to be determined, then the term "power spectrum density" or "power spectrum" should be defined. Both terms are used synonymously as they refer to the Fourier transform of the autocorrelation function or, equivalently, to the square of the amplitude of the Fourier transform of the original data.

In our case, in order to generate acceleration spectra the amplitude Fourier spectrum and the derived power spectral density obtained from the Fourier amplitude spectrum of the observed seismograms are used. The seismic action is defined using deterministic and stochastic seismic response spectra by starting from the information provided by seismograms. The data necessary to develop deterministic response spectra are the maximum ground acceleration and the mean predominant frequency of the earthquake motion. In the case in which stochastic seismic response spectra are simulated, the most important input datum is the power spectral density of the acceleration at the epicentre.

A procedure which allows families of artificial accelerograms which match the proposed response spectra to be generated is also described.

Examples are included to illustrate the procedures described in this paper. Finally the proposed methodology is applied to the seismic zone of the Pyrenees, a typical seismic zone of intermediate seismicity in which the available seismological data are in the form of seismograms.

2 *Methodology for analysing seismograms*

Short-period seismograms, as shown in Fig. 1, can be used to obtain acceleration

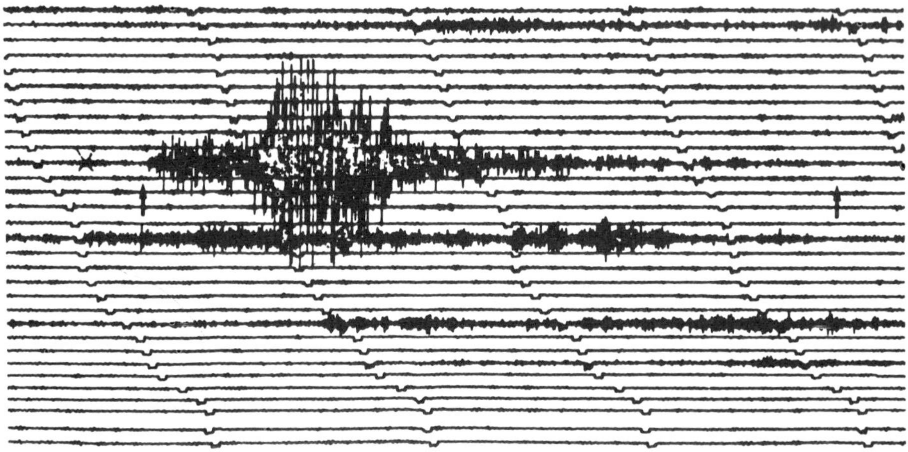

Fig. 1 *Example of a short-period seismogram recorded at the Ebro seismographic station.*

data. For earthquakes of magnitude 6 or below, the most prominent are generally L_g waves. The vertical components of L_g waves are composed of higher modes of Rayleigh waves, and the horizontal components are composed of higher modes of Rayleigh and Love waves. Their propagation velocity is about 3.5 km s^{-1}. In short-period seismograms they normally appear as the largest-amplitude arrivals, immediately before the exponential decay of the seismic signal.

The following important points have to be taken into account during analysis of the data.

(a) To avoid digitizing problems if analogue seismograms are used, it is recommended that the short-period seismograms are digitized at irregular time intervals. *A posteriori* cubic spline interpolation provides the appropriate constant time interval.

(b) To duplicate the original seismogram in detail, special care has to be taken when the peaks of the waveforms or places where the slope changes rapidly are digitized. In such regions a larger number of points must be used.

(c) In order to avoid errors in the amplitude data, departures from the expected amplitude responses must be checked.

(d) If the baseline from which the digital values of the seismograms are measured is correctly selected, an important source of error is avoided (Mitchell and Landismann, 1969; James and Linde, 1971). The method used in this study is to construct a straight line from the point where a trace terminates at one end of the record to the same point in time at the other end of the record.

(e) Only seismograms with clear waveforms should be used in the analysis. In this way it is possible to avoid errors affecting amplitudes and frequencies.

(f) A photographic copy must be made of each of the records used. The digitized record should be plotted and compared with the original. The agreement between the two records must be satisfactory.

(g) The effects due to the different geological structures on which the seismographic stations are built must be considered. In the analysis of seismograms from different sites, special care must be taken regarding the different amplification caused by different crustal structures. A prior investigation of the local amplification factors has to be performed. Sediment thicknesses may play an important role when local conditions are investigated.

When the above considerations have been taken into account, the application of Fourier analysis to digitized or digital seismic waveforms provides the amplitude spectrum of the ground motion at the recording station. Corrections due to the geometrical distribution of energy and anelastic effects yield the amplitude spectrum at the epicentre. When starting from displacement time histories, successive multiplication of the displacement spectrum by ω provides the acceleration spectrum from which the solution to the earthquake engineering problem is obtained.

3 *Stochastic seismic response spectra*

3.1 Stochastic model of the ground acceleration

The acceleration of an earthquake at its epicentre can be defined in a simplified way as a white noise process with a sample function $a(t)$ and a power spectral density S_0. This process is filtered through a model of the soil layer so that the effect of the local soil conditions is included in the analysis. A single-degree-of-freedom model introduced by Kanai and Tajimi (Tajimi, 1960; Penzien, 1965) is used first. Its frequency response function is

$$|H_{1g}(\theta)|^2 = \frac{1+4v_{g1}(\theta/\theta_{g1})^2}{\{1-(\theta/\theta_{g1})^2\}^2+4v_{g1}{}^2(\theta/\theta_{g1})^2} \tag{1}$$

where θ_g is its natural frequency and v_g is its damping ratio. For firm soil conditions Kanai has recommended the values $\theta_g = 15.6\ \text{rad s}^{-1}$ and $v_g = 0.6$.

The power spectral density $S_a^f(\theta)$ obtained after filtering the white noise process through this model is given by

$$S_a^f(\theta) = |H_{1g}(\theta)|^2\, S_0 \qquad \theta \in (-\infty, +\infty) \tag{2}$$

This filter reduces the high frequency components and amplifies those components which have a frequency in the neighbourhood of $\theta = \theta_g$. However, the filter also produces singularities for very low frequencies and consequently unbounded variance of the ground velocities and displacements (Clough and Penzien, 1975). A plot of the power spectral density function $S_a^f(\theta)$ is given in Fig. 2. To avoid this problem the resultant process can be filtered a second time through a filter which

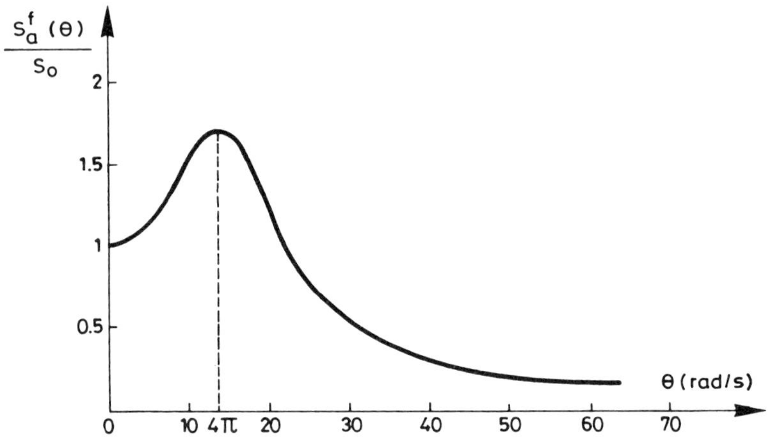

Fig. 2 *Plot of the power spectral density function* $S_a^{\mathrm{f}}(\theta)$.

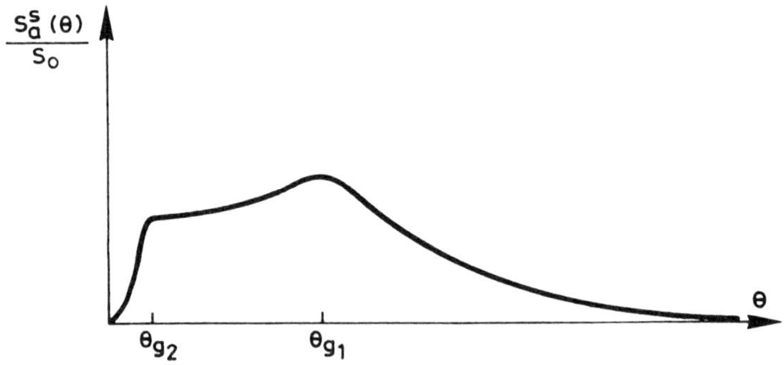

Fig. 3 *Plot of the power spectral density function* $S_a^{\mathrm{s}}(\theta)$.

reduces the very low frequency components. The following transfer function can be used for the second filter (Clough and Penzien, 1975; Barbat, 1982):

$$|H_{2g}(\theta)|^2 = \frac{(\theta/\theta_{g2})^2}{\{1-(\theta/\theta_{g2})^2\}^2 + 4v_{g2}{}^2(\theta/\theta_{g2})^2} \tag{3}$$

The process obtained using both filters has a power spectral density $S_a^{\mathrm{s}}(\theta)$ given by

$$S_a^{\mathrm{s}}(\theta) = |H_{1g}(\theta)|^2 |H_{2g}(\theta)|^2 S_0 = |H_g(\theta)|^2 S_0 \tag{4}$$

Figure 3 shows a plot of the power spectral density function $S_a^{\mathrm{s}}(\theta)$.

The power spectral density of a process which describes seismic ground acceleration is defined by the seismic codes of different countries or different international organizations. For example, the Joint Committee on Structural Safety (CEB–CECM–CIB–FIP–IABSE) has developed power spectral densities for

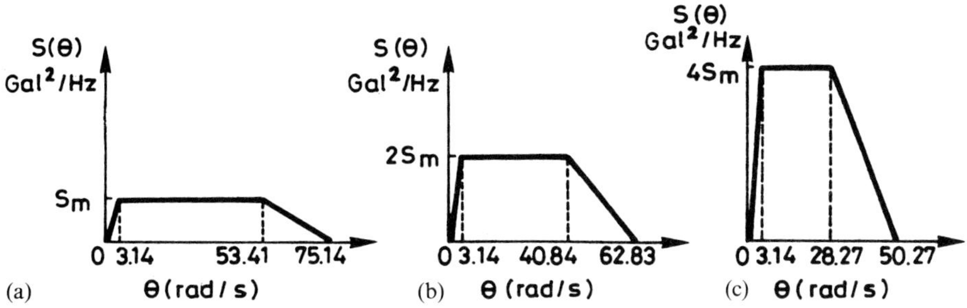

Fig. 4 *Power spectral densities for (a) firm soil, (b) medium soil and (c) soft soil. (After Lungu and Ghiocel, 1982.)*

seismic action expressed as a function of the soil characteristics (Joint Committee on Structural Safety, 1974; Lungu and Ghiocel, 1982). Plots of these power spectral densities are given in Fig. 4 for three different soil conditions: firm soil (Fig. 4(a)), medium soil (Fig. 4(b)) and soft soil (Fig. 4(c)).

Most seismic events have time-dependent characteristics. Therefore it is important to consider the non-stationarity of the random processes used in modelling a seismic ground acceleration. Such processes can be generated in a simplified way by transforming a stationary process $a(t)$ to a non-stationary process $\tilde{a}(t)$ by multiplication by a deterministic time function $I(t)$:

$$\tilde{a}(t) = a(t)\,I(t) \tag{5}$$

Clough and Penzien (1975) show that the intensity functions can be obtained for a given seismic area by using a statistical analysis of actual ground acceleration records obtained in that area. The following intensity function proposed by Oto (see Damrath, 1974) was used in this study to generate non-stationary processes:

$$I(t) = \frac{t}{t_0}\exp\left(1-\frac{t}{t_0}\right) \tag{6}$$

3.2 First-passage problem for a stationary stochastic model

The power spectral density $S_x(\theta)$ of the displacement response of a single-degree-of-freedom model is related to the power spectral density of the ground motion acceleration $S_a(\theta)$ by the equation

$$S_x(\theta) = m^2|H(\theta)|^2 S_a(\theta) \tag{7}$$

where m is the mass of the model, θ is the frequency of the excitation and $H(\theta)$ is the frequency response function of the model. Its modulus is

$$|H(\theta)| = \frac{1}{k\{(1-\theta^2/\omega^2)^2+(2\nu\theta/\omega)^2\}^{1/2}} \tag{8}$$

Equations similar to (7) but relating the power spectral densities of the velocity and acceleration responses to the power spectral density of the ground acceleration can also be obtained:

$$S_{\dot{x}}(\theta) = m^2\theta^2|H(\theta)|^2 S_a(\theta) \tag{9}$$

$$S_{\ddot{x}}(\theta) = m^2\theta^4|H(\theta)|^2 S_a(\theta) \tag{10}$$

The variances of the displacement, velocity and acceleration responses of the structure are given by

$$\sigma_x{}^2 = \int_{-\infty}^{+\infty} S_x(\theta)\,\mathrm{d}\theta \qquad \sigma_{\dot{x}}{}^2 = \int_{-\infty}^{+\infty} S_{\dot{x}}(\theta)\,\mathrm{d}\theta \qquad \sigma_{\ddot{x}}{}^2 = \int_{-\infty}^{+\infty} S_{\ddot{x}}(\theta)\,\mathrm{d}\theta \tag{11}$$

In the case of a ground acceleration modelled as a filtered white noise, the variances of the displacement, velocity and acceleration responses expressed using Eqns (11) are

$$\sigma_x{}^2 = m^2 S_0 \int_{-\infty}^{+\infty} |H(\theta)|^2 |H_g(\theta)|^2\,\mathrm{d}\theta \tag{12}$$

$$\sigma_{\dot{x}}{}^2 = m^2 S_0 \int_{-\infty}^{+\infty} \theta^2 |H(\theta)|^2 |H_g(\theta)|^2\,\mathrm{d}\theta \tag{13}$$

$$\sigma_{\ddot{x}}{}^2 = m^2 S_0 \int_{-\infty}^{+\infty} \theta^4 |H(\theta)|^2 |H_g(\theta)|^2\,\mathrm{d}\theta \tag{14}$$

In general the integrals in these equations can only be solved numerically. In the particular case in which the ground motion acceleration $a(t)$ is described as an ideal white noise, $|H(\theta)|^2 = 1$ and consequently $S_a(\theta) = S_0$. The integrals in Eqns (12), (13) and (14) can then be solved analytically, and the following variances are obtained for the response processes:

$$\sigma_x{}^2 = \frac{\pi S_0}{2\nu\omega^3} \qquad \sigma_{\dot{x}}{}^2 = \frac{\pi S_0}{2\nu\omega} \qquad \sigma_{\ddot{x}}{}^2 = \frac{\pi S_0 \omega}{2\nu} \tag{15}$$

The **extreme mean value** x_e of the structural response process $x(t)$ is defined as the response expected only once during a given finite time interval t_e. The first-passage problem of computing x_e can be solved under some restrictive hypotheses. For example, Damrath (1974) has considered the hypothesis that the process is a narrow band.

The number of cases in which the response overpasses the positive or negative value x_e during the time interval t_e is denoted $N(\pm x_e)$, and $n(\pm x_e)$ is the number of overpasses per time unit. Thus

$$N(\pm x_e) = n(\pm x_e)\,t_e = 1 \tag{16}$$

It can be demonstrated (Ventsel, 1973) that

$$n(+x_e) = \int_0^\infty p(x_e, \dot{x})\dot{x}\,d\dot{x} \tag{17}$$

This equation expresses the probability that $x(t) \in [x_e, x_e + dx]$ and at the same time $\dot{x}(t) \in [\dot{x}, \dot{x} + d\dot{x}]$. The second-order gaussian probability density of the variables x and $\dot{x}$ is written as

$$p(x, \dot{x}) = \frac{1}{2\pi\sigma_x\sigma_{\dot{x}}} \exp\left[-\frac{1}{2}\left\{\left(\frac{x}{\sigma_x}\right)^2 + \left(\frac{\dot{x}}{\sigma_{\dot{x}}}\right)^2\right\}\right] \tag{18}$$

where it has been assumed that $x(t)$ and $\dot{x}(t)$ have zero means. For the extreme mean value x_e the probability density becomes

$$p(x_e, \dot{x}) = \frac{1}{2\pi\sigma_x\sigma_{\dot{x}}} \exp\left[-\frac{1}{2}\left\{\left(\frac{x_e}{\sigma_x}\right)^2 + \left(\frac{\dot{x}}{\sigma_{\dot{x}}}\right)^2\right\}\right] \tag{19}$$

Substitution of (19) in (17) leads to (Damrath, 1974)

$$n(+x_e) = \frac{1}{2\pi}\frac{\sigma_{\dot{x}}}{\sigma_x} \exp\left\{-\frac{1}{2}\left(\frac{x_e}{\sigma_x}\right)^2\right\} \tag{20}$$

If the process $x(t)$ is stationary, the relation

$$\frac{\sigma_{\dot{x}}}{\sigma_x} = \theta_m \tag{21}$$

can be used to define a mean frequency θ_m for the process. Substitution of (21) in (20) gives

$$n(+x_e) = \frac{\theta_m}{2\pi} \exp\left\{-\frac{1}{2}\left(\frac{x_e}{\sigma_x}\right)^2\right\} \tag{22}$$

Since

$$n(\pm x_e) = 2n(+x_e) \tag{23}$$

substitution of (22) in (23) gives

$$n(\pm x_e) = t_e\frac{\theta_m}{\pi} \exp\left\{-\frac{1}{2}\left(\frac{x_e}{\sigma_x}\right)^2\right\} \tag{24}$$

Equation (16) now enables us to establish the following equation which defines x_e:

$$t_e\frac{\theta_m}{\pi} \exp\left\{-\frac{1}{2}\left(\frac{x_e}{\sigma_x}\right)^2\right\} = 1 \tag{25}$$

The solution is

$$x_e = \sigma_x\left\{2\ln\left(t_e\frac{\theta_m}{\pi}\right)\right\}^{1/2} \tag{26}$$

In practice, the value x_e is similar to an ordinate of the mean maximum seismic response spectrum which corresponds to the ground motion acceleration process $a(t)$.

3.3 First-passage problem for a non-stationary stochastic model

For a non-stationary stochastic model an evolutionary power spectral density of the non-stationary response process $\tilde{x}(t)$, denoted $S_{\tilde{x}}(\theta,t)$ must be used. Such a power spectral density has been introduced by Priestley (1965). In this study the power spectral density of the non-stationary response process of the single-degree-of-freedom model is expressed as a function of the stationary power spectral density $S_x(\theta)$ by means of the following equation (Barbat, 1978):

$$S_{\tilde{x}}(\theta,t) = |G_{\tilde{x}}(\theta,t)|^2 S_x(\theta) \tag{27}$$

where $|G_{\tilde{x}}(\theta,t)|^2$ can be interpreted as a complex transfer function of the non-stationary displacement response. This function uses the effect of the intensity function $I(t)$ on the stationary process to simulate a non-stationary process. For a stationary process $|G_{\tilde{x}}(\theta,t)|^2 = 1$. By using the transfer function $H(\theta)$ of the single-degree-of-freedom model the power spectral density of the non-stationary response process can be related directly to the power spectral density of the stationary excitation process $S_a(\theta)$:

$$S_{\tilde{x}}(\theta,t) = m^2|G_{\tilde{x}}(\theta,t)|^2|H(\theta)|^2 S_a(\theta) \tag{28}$$

A similar expression can be found for the power spectral density of the velocity response process $S_{\dot{\tilde{x}}}(\theta,t)$:

$$S_{\dot{\tilde{x}}}(\theta,t) = m^2|G_{\dot{\tilde{x}}}(\theta,t)|^2|H(\theta)|^2 S_a(\theta) \tag{29}$$

where $|G_{\dot{\tilde{x}}}\theta,t)|^2$ is the transfer function of the non-stationary velocity response process. For a stationary process $S_{\dot{\tilde{x}}}(\theta,t) = \theta^2$.

For a non-stationary process with an evolutionary power spectral density the variance of the process is also a function of time. The variances $\{\sigma_{\tilde{x}}(t)\}^2$ and $\{\sigma_{\dot{\tilde{x}}}(t)\}^2$ of the displacement response process and the velocity response process respectively can be calculated using expressions similar to Eqn (11) (Lin, 1967; Damrath, 1974):

$$\{\sigma_{\tilde{x}}(t)\}^2 = \int_{-\infty}^{+\infty} S_{\tilde{x}}(\theta,t)\,\mathrm{d}\theta \tag{30}$$

$$\{\sigma_{\dot{\tilde{x}}}(t)\}^2 = \int_{-\infty}^{+\infty} S_{\dot{\tilde{x}}}(\theta,t)\,\mathrm{d}\theta \tag{31}$$

The integrals in Eqns (30) and (31) must be solved numerically.

In the case in which the response process $\tilde{x}(t)$ is non-stationary, the variances $\sigma_{\dot{\tilde{x}}}$ and $\sigma_{\tilde{x}}$ are only approximately proportional. However, Damrath (1974) has defined a mean frequency of the process given by

$$\theta_m = \frac{\sigma_{\dot{\tilde{x}}}}{\sigma_{\tilde{x}}} \tag{32}$$

The process can be treated as stationary in an infinitesimal time interval dt. Thus, according to (24), the expected number of cases in which $\tilde{x}(t)$ overpasses a given value $\pm\tilde{x}_e$ in the time interval dt is

$$n(\pm\tilde{x}_e) = \frac{\theta_m}{\pi}\exp\left\{-\frac{1}{2}\left(\frac{\tilde{x}_e}{\sigma_{\tilde{x}}}\right)^2\right\}\mathrm{d}t \tag{33}$$

Since in this case

$$N(\pm\tilde{x}_e) = \int_0^\infty n(\pm\tilde{x}_e)\,\mathrm{d}t \tag{34}$$

the equation

$$N(\pm\tilde{x}_e) = 1$$

which defines $\tilde{x}_e$ can be expressed in the following form:

$$\frac{\theta_m}{\pi}\int_0^\infty \exp\left\{-\frac{1}{2}\left(\frac{\tilde{x}_e}{\sigma_{\tilde{x}}}\right)^2\right\}\mathrm{d}t = 1 \tag{35}$$

Equation (35) can only be solved for x_e numerically, but it can be simplified if it is assumed that the following equation holds:

$$\sigma_{\tilde{x}}(t) = \sigma_{\tilde{x}_m}\frac{t}{t_{\tilde{x}_m}}\exp\left(1-\frac{t}{t_{\tilde{x}_m}}\right) \tag{36}$$

where $\sigma_{\tilde{x}_m}$ is the maximum standard deviation and $t_{\tilde{x}_m}$ is the time at which the maximum standard deviation occurs. If

$$z = \frac{t}{t_{\tilde{x}_m}} \tag{37}$$

is now substituted in Eqn (36), it takes the form

$$\sigma_{\tilde{x}}(z) = \sigma_{\tilde{x}_m} z\exp(1-z) \tag{38}$$

and Eqn (35) can be expressed as

$$\frac{\theta_m}{\pi}\int_0^\infty \exp\left[-\frac{1}{2}\left\{\frac{\tilde{x}_e}{\sigma_{\tilde{x}_m} z\exp(1-z)}\right\}^2\right]t_{\tilde{x}_m}\,\mathrm{d}z = 1 \tag{39}$$

The time $t_{\tilde{x}_m}$ is calculated using the formula

$$t_{\tilde{x}_m} = \tfrac{1}{2}(t_{G_m} + t_0) \tag{40}$$

where t_0 and t_{G_m} are respectively the times at which the intensity function $I(t)$ and the function $G_{\tilde{x}}(\omega_v, t)$ take a maximum value. t_{G_m} is obtained by using a stepwise procedure. Thus the maximum standard deviations of $\tilde{x}$ and $\dot{\tilde{x}}$ can be calculated as follows:

$$\sigma_{\tilde{x}_m} = \sigma_{\tilde{x}}(t_{\tilde{x}_m}) \tag{41}$$

$$\sigma_{\dot{\tilde{x}}_m} = \sigma_{\dot{\tilde{x}}}(t_{x_m}) \tag{42}$$

Equation (38) is now expressed in the form

$$\int_0^\infty \exp\left[-\frac{1}{2}\left\{\frac{\tilde{x}_e}{\sigma_{\tilde{x}_m}}\frac{1}{z\exp(1-z)}\right\}^2\right]\mathrm{d}z = \frac{\pi}{\theta_m t_{\tilde{x}_m}} \tag{43}$$

and the value

$$\alpha = \frac{\tilde{x}_e}{\sigma_{\tilde{x}_m}} \tag{44}$$

is obtained numerically. Finally

$$\tilde{x}_e = \alpha\sigma_{\tilde{x}_m} \tag{45}$$

is obtained as a solution of the problem. It represents the ordinate of a response spectrum corresponding to a stochastic non-stationary model of the ground acceleration.

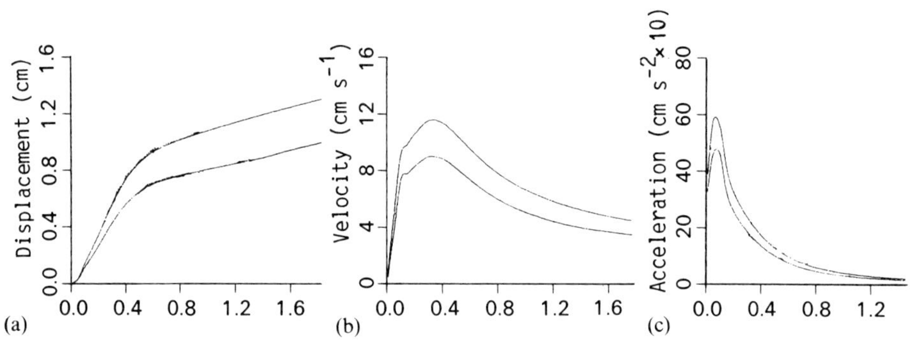

Fig. 5 *Comparison of response spectra for stationary and non-stationary processes: (a) displacement spectra; (b) velocity spectra; (c) acceleration spectra.*

Example Comparison of simulated stationary and non-stationary stochastic response spectra

Consider a process with a power spectral density of 1500 Gal^2 Hz^{-1} at the epicentre. Compute and compare the seismic response spectra obtained for the cases that the input is (a) a stationary process and (b) a non-stationary process. The damping ratio v of the model is 0.2. Filters $|H_{1g}|^2$ and $|H_{2g}|^2$ are both used. The following values are assumed for the filter characteristics: $\theta_{g1} = 15.6$ rad s^{-1}, $v_{g1} = 0.6$, $\theta_{g2} = 100$ rad s^{-1} and $v_{g2} = 0.3$.

The results of the computation are shown in Fig. 5. The displacement response spectra for stationary and non-stationary processes are compared in Fig. 5(a), and the same comparison is performed for the velocity and acceleration spectra in Figs 5(b) and 5(c) respectively.

4 Deterministic estimation of seismic response spectra

For seismic zones for which only seismograms are available, it is necessary to obtain approximate values of the maximum spectral accelerations and also to investigate the frequency content of the seismic signal. These data enable an approximate definition to be made of the seismic response spectra to be used in the structural response computations.

Canas (1986) has developed a formula for calculating pseudo-acceleration response spectra ordinates in the Pyrenees region. As pointed out by Canas, this kind of formula should be developed for the particular seismic zone in which response spectra have to be estimated and used to calculate the maximum ground acceleration. It must be emphasized that the formulae, which are useful for zones of low seismicity, are approximate owing to the incomplete instrumental data generally available in such cases. Moreover, they do not provide the complete information necessary to obtain seismic spectra, as these require the definition of

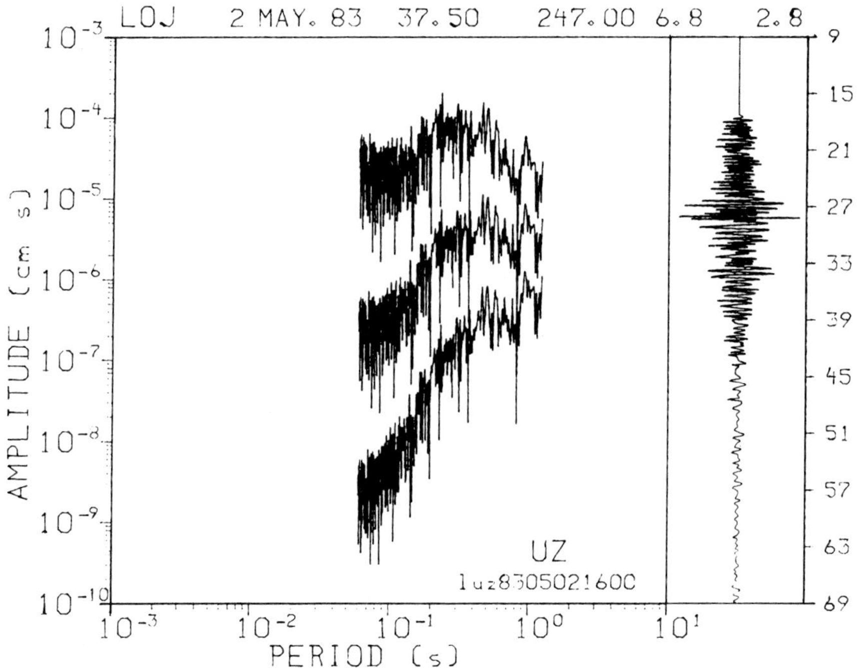

Fig. 6 Example of displacement (bottom), velocity (middle) and acceleration (top) Fourier amplitude spectra of the digitized seismogram shown on the right. The seismogram was recorded at the Cartuja Observatory, Spain.

spectral ordinates for each frequency. The frequency content of the ground motion can be studied using the Fourier amplitude spectra obtained by starting from available seismograms like that shown in Fig. 6. The predominant frequency can be obtained by analysing the envelope of the amplitude spectrum.

We now briefly review the methodology used to obtain pseudo-acceleration data for regions without accelerograms.

First it is necessary to determine the anelastic attenuation coefficient γ associated with the region of study. This can be done by using either coda waves or L_g waves.

Coda waves appear in seismograms as waves with an exponential decrease arriving after the L_g waves. The peak-to-peak amplitude of coda waves can be expressed as

$$\frac{A(t^*)}{\sqrt{8}} = Q^{1/2} M_0 B(f_p) C(f_p, t^*) \tag{46}$$

where $t^* = t/Q$ (t is the elapsed time between the time of origin of the earthquake and observation of the predominant frequency on the seismogram and Q is the anelastic attenuation factor), M_0 is the seismic moment and $B(f_p)$ is the coda excitation factor. The coda shape function $C(f_p, t^*)$ is given by

$$C(f_p, t^*) = I(f_p)(t^*)^{-1/2} \left(\frac{\mathrm{d}f_p}{\mathrm{d}t^*}\right)^{1/4} \exp(-\pi f_p t^*) \tag{47}$$

where $I(f_p)$ is the instrument magnification for the predominant frequency f_p. The Q values are obtained by matching the theoretical f_p–t^* curves corresponding to the seismographic stations used in the study with the observed f_p–t data determined for each station. The anelastic attenuation coefficients γ can be determined using the well-known expression (Nuttli, 1973)

$$\gamma = \frac{\pi}{QUT} \tag{48}$$

where U is the group velocity of the L_g waves and T is the period. The γ value obtained using this method is one approach to the true γ value of the L_g waves. If there are enough L_g amplitude data, it is recommended that the γ values are determined using the maximum sustained amplitude of L_g waves by means of the expression

$$A = A_0 r^{-5/6} \exp(-\gamma r) \tag{49}$$

where A is the maximum sustained amplitude, A_0 is the amplitude at the epicentre and r is the epicentral distance.

If the attenuation effect is known, the observed amplitude Fourier spectrum at the sites can be corrected using the anelastic attenuation of the ground. The use of this correction together with the correction due to geometrical spreading of the energy will yield the amplitude spectra at the epicentre.

Secondly, the spectral displacement amplitudes f_d given by the Fourier transform have the form

$$f_d = \left| \int_{-\infty}^{\infty} f(t) \exp(-\mathrm{i}\,\omega t)\,\mathrm{d}t \right| \tag{50}$$

where $f(t)$ is the digitized seismic signal, t is the time and ω is the angular frequency. The spectral acceleration can be obtained from the expression

$$f_s = \omega^2 f_d \tag{51}$$

If we start from the spectral velocity amplitudes f_v, then f_s is given by

$$f_s = \omega f_v \tag{52}$$

For engineering purposes it is useful to work with pseudo-parameters. The approximate relationship between the pseudo-velocities PSV and the Fourier spectral acceleration is given by

$$\text{PSV} \approx f_s \tag{53}$$

since in PSV the damping coefficient is zero. Therefore the pseudo-acceleration PSA can be written as

$$\text{PSA} = \omega\,\text{PSV} \tag{54}$$

The peak values and predominant frequencies of the pseudo-acceleration spectra can now be obtained using Eqn (54).

Experience has shown that the peak acceleration can be expressed as a function of the magnitude m, the anelastic attenuation coefficient γ and the epicentral distance r:

$$\text{PSA} = A \exp(Bm) \exp(-\gamma r)\, r^{-1/2} \tag{55}$$

where A and B are constants to be determined. Local pseudo-acceleration formulae that may be important for regions without acceleration data can be determined by applying the least-squares method to a set of earthquakes with known magnitudes and locations recorded at one or more seismographic stations and covering a specific area with a known anelastic attenuation coefficient.

A bilinear acceleration response spectrum is thus estimated using the defined maximum acceleration and the predominant frequency.

5 Generation of artificial accelerograms which match a seismic response spectrum

Families of artificial accelerograms can be generated by starting from simulated stochastic response spectra. The procedure described below is due to Gasparini and Vanmarke (1976). It starts from the expansion of a periodic function in a series of sinusoidal waves of the form

$$x(t) = \sum_{i=1}^{n} A_i \sin(\theta_i t + \phi_i) \tag{56}$$

where A_i is the amplitude, θ_i is the frequency and ϕ_i is the phase angle of the ith sinusoid. Different motions can be obtained by using a random process to generate phase angles with a uniform distribution between 0 and 2π and amplitudes within a given range. The amplitudes A_i are related to the power spectral density $S(\theta_i)$ by

$$S(\theta_i)\Delta\theta = \frac{A_i^2}{2} \tag{57}$$

where $\Delta\theta$ are intervals at which the frequencies θ_i are given. Thus a power spectral density function $S(\theta)$ can be defined which expresses the relative influence of the individual sinusoids on the total power of the motion.

The motion simulated using Eqn (56) is a sample of a stationary process. Intensity functions $I(t)$ like those described by Eqn (6) can be used to represent the transient character of actual earthquakes. A ground motion modelled as a non-stationary process is obtained:

$$z(t) = I(t) \sum_{i=1}^{n} A_i \sin(\theta_i t + \phi_i) \tag{58}$$

The resultant ground motion is stationary with respect to frequency. This procedure is used to reduce or increase the maximum acceleration generated until it corresponds to a given value (Gasparini and Vanmarke, 1976).

After the ground motion has been simulated by means of Eqn (58), the corresponding response spectra are calculated and compared with a given target velocity response spectrum. The comparison is performed for a number of prescribed frequencies. The ratio of the target velocity response spectrum ordinate S_v^t to the computed velocity response spectrum ordinate S_v^k, where k is the identifier

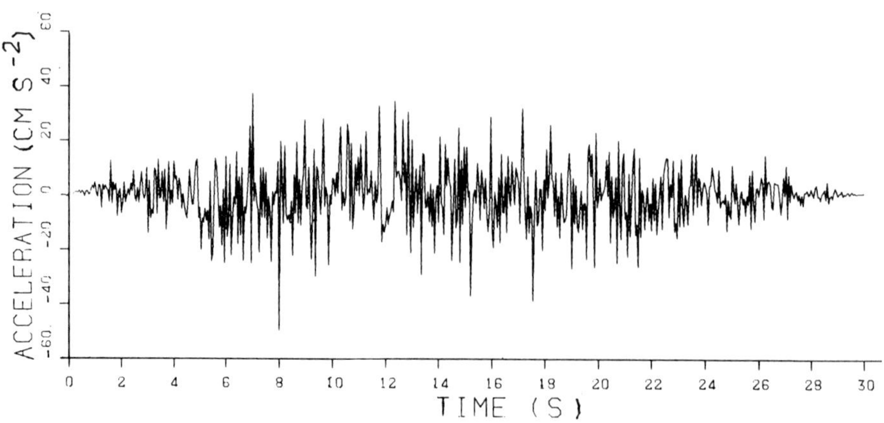

Fig. 7 *Simulated accelerogram.*

of the iteration cycle, is calculated for each control frequency:

$$r_{j,k} = \frac{S_v^{\,t}(\omega_j)}{S_v^{\,k}(\omega_j)} \tag{59}$$

The power spectral density of the simulated motion is modified by $r_{j,k}^{\,2}$:

$$S^{k+1}(\omega_j) = S^k(\omega_j) \times r_{j,k}^{\,2} \tag{60}$$

In the next iteration cycle a new simulation of the ground motion is performed using the modified power spectral density and a new response spectrum is calculated.

The procedure does not ensure convergence for all control frequencies because the response for a particular frequency also depends on the values of the frequencies near to it. Generally, convergence is achieved in about four iteration cycles.

Example Simulation of an accelerogram which matches a stochastic seismic response spectrum

A simulated accelerogram which matches the stochastic non-stationary velocity response spectrum is shown in Fig. 7. The program SIMQKE (Gasparini and Vanmarke, 1976) has been used in the simulation.

6 *Application to the seismic zone of the Pyrenees*

The method described in Section 4 has been applied to seismograms corresponding to earthquakes located in or near the Pyrenees (Table 1) and recorded at the Ebro seismographic station (Fig. 8). All the earthquakes have a duration magnitude between 2.8 and 5.6 (Peñuelas, 1985).

The aims of this part of the study were (a) to determine a peak acceleration formula for the region under investigation, (b) to determine the approximate predominant frequencies of the pseudo-accelerations and (c) to obtain an estimated acceleration pseudo-response spectrum and the corresponding velocity pseudo-response spectrum.

The predominant frequencies and pseudo-acceleration data were obtained by applying the method to the seismograms listed in Table 1. The pseudo-acceleration formula for the region of interest is

$$\text{PSA} = -1.980(\pm 0.986) + 0.880(\pm 0.211)m - 0.5\log_{10} r - 0.013(\log_{10} e)r \tag{61}$$

Results obtained from expression (61) for various magnitudes and epicentral distances are shown in Fig. 9. Expression (61) was obtained using a short-period instrument (vertical component) with similar characteristics to those of the WWSSN system. Therefore a refined formula may be obtained if a higher frequency instrument than the one used here becomes available. Owing to the lack of acceleration data in the region under consideration, expression (61) can be

Table 1

Date	*Time of origin*	*Depth* (km)	*Location*	
			Longitude	*Latitude*
1/03/76	02:26:03.3	1	1°24.4 E	41°27.3 N
3/06/76	15:25:18.4	3	1°22.7 E	42°23.7 N
10/06/76	03:24:08.2	1	1°26.1 E	42°24.5 N
11/06/76	11:12:09.3	2	1°27.2 E	42°24.9 N
11/06/76	21:53:05.9	5	1°27.9 E	42°35.6 N
11/06/76	21:56:45.4	5	0°55.8 E	42°14.0 N
12/06/76	04:18:03.5	5	1°21.8 E	42°33.6 N
1/04/79	19:04:29.8	10	2°15.4 W	43°09.0 N
25/05/79	01:42:59.8	—	2°48.0 E	42°00.0 N
25/05/79	01:47:12.3	—	2°42.0 E	42°00.0 N
25/05/79	07:32:03.4	—	2°42.0 E	41°54.0 N
1/07/79	10:47:12.8	5	2°00.0 E	42°42.0 N
6/06/79	11:59:52.4	5	0°16.6 W	43°21.4 N
18/08/79	00:36:06.2	15	0°45.4 E	43°13.2 N
10/10/79	16:18:50.5	20	0°06.2 W	41°18.9 N
5/12/79	23:02:34.1	40	2°24.3 E	42°33.3 N
17/12/79	17:37:15.2	10	0°30.0 E	43°24.0 N
28/01/80	09:25:38.3	10	2°25.1 E	42°32.6 N
2/02/80	05:26:32.0	5	1°05.6 W	41°14.5 N
29/02/80	20:40:50.5	5	0°21.5 W	43°11.1 N
29/02/80	22:58:03.8	5	0°17.7 W	43°16.6 N
29/03/80	22:27:30.3	—	2°24.0 E	42°24.0 N
16/04/80	13:10:27.8	5	0°52.5 W	42°29.8 N
17/05/80	01:14:22.3	5	2°10.3 E	42°34.4 N
19/07/80	23:27:54.0	5	0°06.2 W	42°32.9 N
20/10/80	00:30:35.1	10	0°55.1 W	42°22.8 N
14/12/80	19:06:01.4	5	2°26.2 E	41°54.8 N
10/01/81	18:15:45.8	1	0°06.0 E	40°39.0 N
19/03/81	15:05:16.7	—	1°24.0 W	42°36.0 N
23/04/81	15:53:05.9	5	1°49.0 E	42°50.0 N
23/06/81	15:53:05.9	5	1°49.0 E	42°50.0 N
6/01/82	16:32:49.8	10	0°58.0 W	43°15.0 N
6/01/82	17:22:11.8	18	0°59.0 W	43°14.0 N
11/03/82	14:49:37.7	7	1°54.0 W	42°53.0 N
11/04/82	20:21:46.7	7	3°11.0 E	41°34.0 N
12/04/82	16:30:25.4	14	3°20.0 E	41°23.0 N
7/05/82	11:41:49.8	5	1°42.0 E	41°06.0 N
22/05/82	04:03:03.2	10	1°46.0 W	42°45.0 N
22/05/82	04:37:51.0	5	1°45.0 W	42°51.0 N
11/06/82	16:16:28.6	5	0°38.0 E	41°59.0 N
12/06/82	09:59:15.8	14	0°21.0 W	43°12.0 N
12/06/82	17:23:52.6	5	0°24.0 W	43°15.0 N
22/06/82	19:50:23.7	7	1°52.0 W	42°52.0 N
22/06/82	20:24:05.5	5	1°55.0 W	43°01.1 N
24/12/82	00:06:16.8	5	0°34.0 E	42°34.0 N

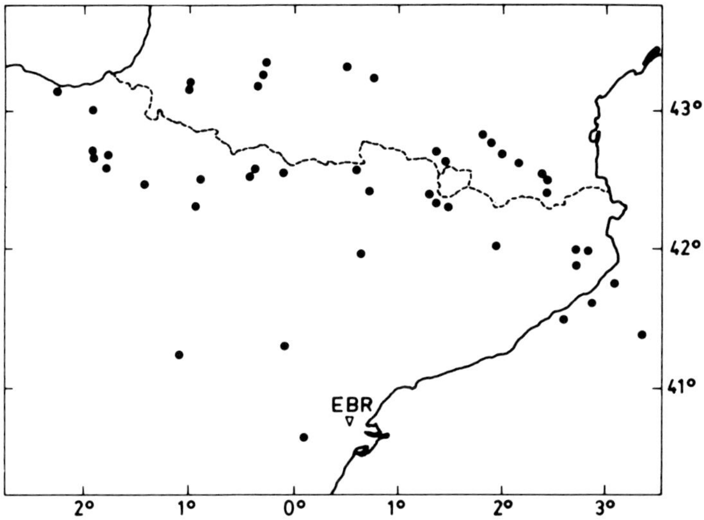

Fig. 8 *Location of the earthquakes and of the Ebro seismographic station (EBR).*

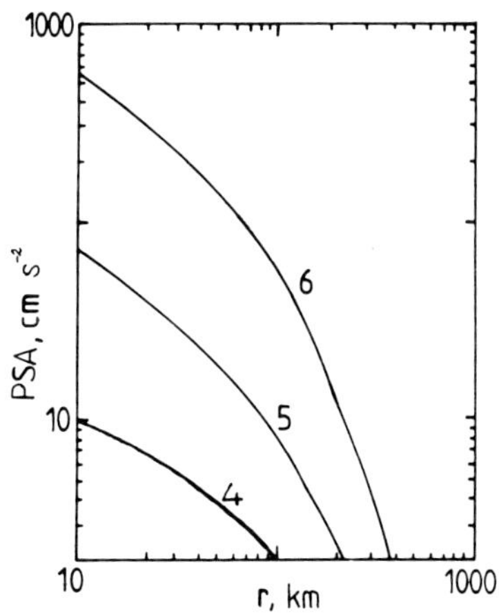

Fig. 9 *Peak values of pseudo-acceleration spectra for various magnitudes and epicentral distances.*

considered as a satisfactory approach to the real accelerations expected in the region.

Figure 10 shows the velocity pseudo-response spectrum obtained for the Pyrenees region by applying the methodology proposed here. An accelerogram matching this pseudo-response spectrum is given in Fig. 11.

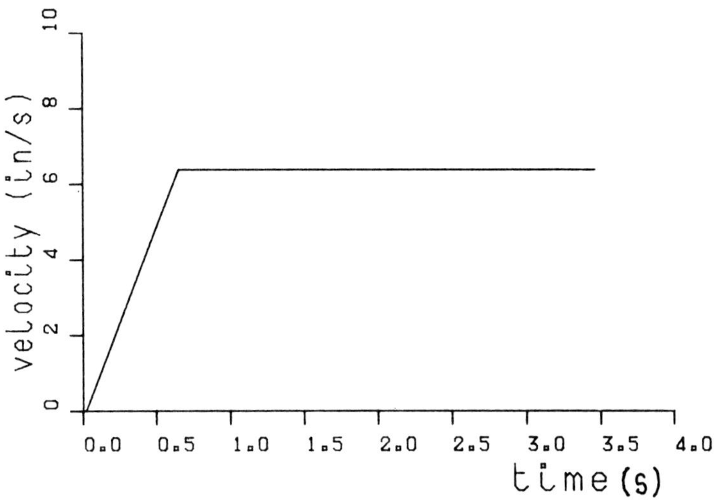

Fig. 10 *Velocity pseudo-response spectrum for the Pyrenees region.*

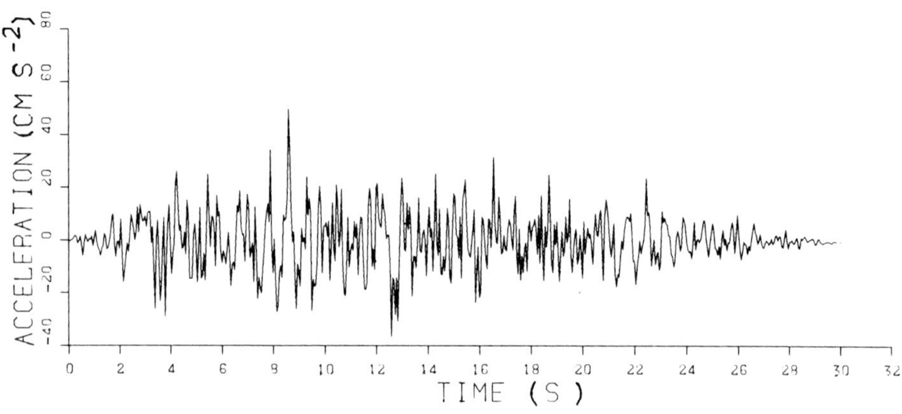

Fig. 11 *Accelerogram matching the pseudo-response spectrum of Fig. 10.*

7 Discussion and conclusions

Simulation of earthquake motion at a site is used in the time-domain computation of the seismic response of structures when actual records of seismic ground motions are not available. The procedures generally used to simulate seismic motion generate artificial accelerograms starting from models based on the theory of the stochastic process. In this paper seismic action is defined through both deterministic and stochastic response spectra, which are obtained by starting from the information provided by seismograms. By using spectral analysis, it has been possible to obtain a formula which relates the peak values of the pseudo-acceleration spectra to magnitude, anelastic attenuation and epicentral distance. The methodology used to obtain this formula may be useful for obtaining similar formulae in regions where acceleration data are poor.

Definition of the seismic response by means of deterministically estimated response spectra has the following advantages.

(1) The starting data necessary for the simulation can be obtained from seismograms.

(2) The seismic response spectra are obtained in a simple way. Nevertheless they are reliable because of the conservative manner in which the maximum acceleration is computed and the fact that maximum ordinates are considered in the high frequency zone.

(3) The response spectra can be adapted to the local soil conditions by using existing procedures.

Stochastic simulation of the response spectra at a given site has the following advantages.

(1) The input data are simple and can be obtained from seismograms.

(2) The simulated spectra are adapted to the ground conditions at a given site in a simple way. This operation is performed by using filters to model the characteristics of the local soil layers. Two filter types have been included in the formulation proposed here, but it is possible to use almost any number of filter types. Obviously, any number of filters of the same type can be included in the simulation if data on the characteristics of these filters are available.

(3) In the procedure proposed here the ground acceleration in the epicentre is defined as white noise. Nevertheless this definition is not essential for the validity of the method. Transfer functions yielding accelerations at the epicentre which are compatible with different faulting models can easily be incorporated in the model. In this way deterministic models which take source characteristics into account can be combined with stochastic simulations to simulate ground motion at a given site.

Both procedures can be completed with any existing method to generate artificial accelerograms which match the stochastic spectra.

The methodology described in this work and applied to the Pyrenees region may be of interest for other regions of the world without acceleration records.

Acknowledgments

This research has been supported in part by the Geological Service of the Generalitat de Cataluña and by the CAICYT (project 1151/84).

References

BARBAT A. H. (1978): The seismic analysis of elevated water tanks considering the interaction phenomena, *Thesis*, Polytechnic Institute of Iassy, Romania.

BARBAT A. H. (1982): *Cálculo Sísmico de Estructuras*, Técnicos Asociados, Barcelona.

CANAS J. A. (1986): Formulas de aceleraciones máximas del terreno para Cataluña. In *Peligrosidad Sísmica de Cataluña*, Report for Generalitat de Cataluña, Barcelona.

CLOUGH R. W. and PENZIEN J. (1975): *Dynamics of Structures*, McGraw-Hill, New York.

DAMRATH R. (1974): Ein Stochastischer Verfahren der Seismischen Analyse von Bauwerken mit Anwendung auf Kernkraftwerke, *Thesis*, Berlin.

GASPARINI D. and VANMARKE E. H. (1976): Simulated earthquake motions compatible with prescribed response spectra, *Report R76-4*, Department of Civil Engineering, Massachusetts Institute of Technology, Cambridge, MA.

JAMES D. E. and LINDE A. T. (1971): A source of major error in the digital analysis of World Wide Standard Station seismograms, *Bulletin of the Seismological Society of America*, **61**, 723–728.

JOINT COMMITTEE ON STRUCTURAL SAFETY (CEB–CECM–CIB–FIP–IABSE) (1974): *Basic Data on Loads* (2nd draft), London.

LIN Y. K. (1967): *Probabilistic Theory of Structural Dynamics*, McGraw-Hill, New York.

LUNGU D. and GHIOCEL D. (1982): *Metode Probabilistice in Calculul Constructiilor*, Tecnica, Bucarest.

MITCHELL B. J. and LANDISMANN M. (1969): Electrical and seismic properties of the earth in southwestern Great Plains of the United States, *Geophysics*, **36**, 363–381.

NUTTLI O. W. (1973): Seismic wave attenuation and magnitude relations for eastern North America, *Journal of Geophysics Research*, **78**, 876–885.

PEÑUELAS G. (1985): *Fórmulas de Magnitud de Duración y Atenuación Anelástica de la Fase L_g para la Subplaca Ibero–Costero–Pirenaica*, Tesina de Especialidad, ETSICCP, UPC, Barcelona.

PENZIEN J. (1965): Application of random vibration theory in earthquake engineering, *Bulletin of the International Institute of Seismology and Earthquake Engineering*, **2**, 47–69.

PRIESTLEY M. B. (1965): Some notes on the physical interpretation of spectra of nonstationary stochastic process, *Journal of Sound and Vibration*, **17** (1), 51–54.

TAJIMI H. (1960): A statistical method of determining the maximum response of building structures during an earthquake, *Proceedings of the 2nd World Conference on Earthquake Engineering, Tokyo*, Vol. 2, pp. 781–798, Science Council of Japan, Tokyo.

VENTSEL H. (1973): *Theorie de Probabilité*, Mir, Moscow.

Pattern Recognition of the Macroseismic Significance of the Parameters of Strong Earth Tremors

A. D. GVISHIANI, A. G. TUMARKIN AND M. N. ZHIZHIN

Institute of the Physics of the Earth, Academy of Sciences of the USSR, Boulevard Gruzinskaya 10, Moscow 123810, USSR

V. SCHENK

Geophysical Institute, Czechoslovak Academy of Sciences, Bocnii 11, Prague 14131, Czechoslovakia

Abstract We report the results of a statistical investigation of the macroseismic significance of the parameters of strong earth tremors. We applied discriminant analysis and the deterministic pattern-recognition algorithm Cora-3 to accelerograms of Californian earthquakes recorded between 1933 and 1971. The accelerograms were classified into three groups depending on the intensity of the earthquake on the Modified Mercalli Scale: slight ($I < 6$); moderate ($I = 6$); destructive ($I > 7$). The results of the statistical analysis and the application of the pattern-recognition algorithm showed that the most significant parameters were the spectral energies at low frequency.

1 Statement of the problem

In this paper we continue our studies (Schenk, 1985; Gvishiani, 1986, 1987) of the following problem in engineering seismology: which values of various parameters of strong earth tremors characterize accelerograms with common macroseismic effects at the location of the accelerograph?

We have attempted to answer this question by applying statistical methods (discriminant analysis) and the deterministic pattern-recognition algorithm Cora-3. In this paper we report the results of this analysis which has enabled us to separate seismic records into three classes on the basis of the macroseismic information furnished by the strong motion parameters. This classification is in good agreement with the observed values of macroseismic intensity and the characteristic features of slight ($I \leqslant 5$) and destructive ($I \geqslant 7$) earthquakes (intensities measured on the Modified Mercalli Scale).

We used 174 accelerograms of Californian earthquakes recorded in the open or in the basements of buildings during the period from 1933 to 1971 and published by the California Institute of Technology (1969–1976). All records were divided into three classes depending on the Mercalli intensity at the location of the accelerograph: slight ($I < 6$); moderate ($I = 6$); destructive ($I > 6$). These classes were chosen as in previous work (Gvishiani *et al.*, 1986, 1987) in order to make the number of records in each class statistically significant. In addition, the threshold $I = 6$ is a natural choice as significant destruction is observed for Mercalli intensities in excess of this value.

Characteristics such as peak and root mean square (r.m.s.) acceleration, energy and impulse, spectral energies at various frequency intervals and the properties of the envelope were chosen as the parameters of strong earth tremors. These parameters are defined formally in Section 2 and the results of the statistical classification and the application of the Cora-3 algorithm are given in Sections 3 and 4. The results show that the most significant parameters are the spectral energies at low frequency.

2 *Description of the parameters*

We performed the discriminant analysis separately for the parameters of the vertical and horizontal components of the accelerograms.

We considered the following 16 parameters for the vertical component $a_z(t)$:

(1) the peak vertical acceleration AVERT;

(2) the "energy" EVERT where

$$\mathrm{EVERT} = \int_{T_0}^{T_1} \{a_z(t)\}^2 \, \mathrm{d}t$$

and T_0 and T_1 are the times at which the vertical component of the acceleration starts and finishes;

(3) $\mathrm{RMSVERT} = \{\mathrm{EVERT}/(T_1 - T_0)\}^{1/2}$

(4) the "impulse" IMPVERT where

$$\mathrm{IMPVERT} = \int_{T_0}^{T_1} |a_z(t)| \, \mathrm{d}t$$

(5)–(14)

$$\mathrm{SPVERT}(N) = \int_{w_{N-1}}^{w_N} |A_z(w)|^2 \, \mathrm{d}w$$

where $A_z(w)$ is the spectrum of $a_z(t)$, and $w_0 = 0.25$ Hz, $w_1 = 0.40$ Hz, $w_2 = 0.63$ Hz, $w_3 = 1.00$ Hz, $w_4 = 1.58$ Hz, $w_5 = 2.50$ Hz, $w_6 = 3.96$ Hz, $w_7 = 6.28$ Hz, $w_8 = 9.95$ Hz, $w_9 = 15.77$ Hz and $w_{10} = 25.00$ Hz, so that $\log_{10}(w_N) - \log_{10}(w_{N-1})$ is constant (the step is uniform on the logarithmic scale);

(15), (16) LRVERT and MRVERT which are related to the shape of the envelope of the vertical acceleration.

First we considered values of the duration DVERT(P) for amplitudes exceeding P% of AVERT ($N = 5, 10, 20, 40, 60, 80, 90, 95$). It is known from studies of the amplitude–frequency distribution of strong tremors (Schenk, 1985) that $\log_{10}$ DVERT(P) depends almost linearly on P. Thus we can use the linear regression

$$\log_{10}\mathrm{DVERT}(P) - \log_{10}\mathrm{DVERT}(5) = -\frac{\mathrm{AVERT}}{\mathrm{LRVERT}}\frac{P}{100} + \frac{\mathrm{MRVERT}}{\mathrm{LRVERT}}$$

with parameters LRVERT and MRVERT.

In the case of the two horizontal components $a_x(t)$ and $a_y(t)$ we considered the following 22 parameters:

(1) $$\mathrm{AHOR} = \max_t \{a_x(t)^2 + a_y(t)^2\}^{1/2}$$

(2) $$\mathrm{EHOR} = \int_{T_0}^{T_1} \{a_x(t)^2 + a_y(t)^2\}\,\mathrm{d}t$$

(3) $$\mathrm{RMSHOR} = \left(\frac{\mathrm{EHOR}}{T_1 - T_0}\right)^{1/2}$$

(4) $$\mathrm{IMPHOR} = \int_{T_0}^{T_1} \{a_x(t)^2 + a_y(t)^2\}^{1/2}\,\mathrm{d}t$$

(5)–(14)

$$\mathrm{SPHOR}(N) = \int_{w_{N-1}}^{w_N} \{|A_x(w)|^2 + |A_y(w)|^2\}\,\mathrm{d}w$$

where the w_N take the same values as in the vertical case;

(15), (16) the linear regression parameters LRHOR and MRHOR which are obtained from durations with total horizontal amplitudes $\{a_x(t)^2 + a_y(t)^2\}^{1/2}$ exceeding P% of AHOR;

(17), (18) the maximum values LRHOR-M and MRHOR-M of the corresponding parameters of each of the horizontal components;

(19) the maximum MHA of the peak horizontal acceleration;

(20) the maximum energy MHE;

(21) the maximum r.m.s. acceleration MHRMS;

(22) the maximum impulse MHIMP.

Table 1 *Statistical significance of the logarithms of the vertical parameters for the problem of classification into three macroseismic classes*

Parameter	*Correlation coefficient with I*	*F statistic*	*Selected in step*	*Discarded in step*
SPVERT(4)	0.72694	92.838	1	
SPVERT(3)	0.70978	89.784		
SPVERT(2)	0.65788	71.258	2	
EVERT	0.72236	69.452		
IMPVERT	0.69961	67.725	5	
SPVERT(5)	0.68763	65.046		
SPVERT(6)	0.69459	60.137		
AVERT	0.68246	54.476	3	
RMSVERT	0.66703	52.121	4	
SPVERT(1)	0.57644	48.492		
SPVERT(7)	0.65784	48.408		
SPVERT(8)	0.65566	47.735	6	
SPVERT(9)	0.58676	35.470		
SPVERT(10)	0.57216	32.721		
LRVERT	0.55129	16.480		
MRVERT	−0.11208	10.556		

Table 2 *Statistical significance of the logarithms of the horizontal parameters for the problem of classification into three macroseismic classes*

Parameter	*Correlation coefficient with I*	*F statistic*	*Selected in step*	*Discarded in step*
EHOR	0.74497	83.483	1	4
MHE	0.73972	81.593		
SPHOR(3)	0.69653	81.375		
SPHOR(4)	0.70334	76.981		
SPHOR(1)	0.63906	73.018	2	
SPHOR(5)	0.71657	71.855		
SPHOR(2)	0.66343	70.615		
IMPHOR	0.70199	68.852		
MHIMP	0.69718	67.407		
SPHOR(6)	0.71629	67.321		
MHA	0.69491	66.053		
AHOR	0.69328	65.876	3	
RMSHOR	0.69879	63.879		
MHRMS	0.69335	62.448	5	
SPHOR(7)	0.67152	54.414		
SPHOR(8)	0.62673	46.274		
SPHOR(9)	0.61180	42.966		
SPHOR(10)	0.62749	41.925		
LRHOR-M	0.57890	17.251		
LRHOR	0.48256	12.557		
MRHOR	−0.31165	6.994	6	
MRHOR-M	−0.27757	5.187		

The first columns of Tables 1 and 2 show the correlation coefficients of the logarithms (base 10) of all these parameters with the intensity I.

Logarithms to the base 10 of the parameters under consideration are used so that an approximately normal distribution is obtained for each parameter, thus enabling discriminant analysis algorithms to be used.

3 Statistical classification of records into three macroseismic classes

Discriminant analysis deals with the following type of problem. Assume that there are a number of distinct classes of objects (in this case macroseismic classes of records of strong tremors) and that each class is represented by a sample of observations. The problem is to find a rule based on given samples which enables an object to be assigned to the correct class on the basis of its parameters. This is done by constructing discriminant functions for each class and assigning the object to the class for which its discriminant function has the largest value, i.e. the class to which it has the highest probability of belonging. If the vectors of the parameters of observations in each class have a multivariate normal distribution, the discriminant functions are linear if the covariance matrices of the classes are identical and quadratic if they are not identical.

First we used stepwise discriminant analysis based on estimation of the F statistics of selection and rejection of parameters in order to identify the most significant parameter, the most significant pair of parameters etc. A description of the algorithm can be found in Enslein *et al.* (1977). Results for the vertical and horizontal cases are presented in Tables 1 and 2 respectively. The values of the F statistics which characterize the information content of each parameter for the problem under consideration are given in the second column of these tables. The F statistic provides a quantitative measure for distinguishing the mean values of the parameter for various classes. Thus in the first step we select the parameter with the highest value of the F statistic, in the second step we add the parameter which gives maximum additional information that will enable us to distinguish the classes in terms of the corresponding F statistic of selection and so on. The order in which parameters are selected and discarded according to this procedure is given in the third and fourth columns of Tables 1 and 2.

It is found that the most important vertical parameter is SPVERT(4) and that SPVERT(2) provides the maximum additional information. EHOR is the most significant horizontal parameter and the best pair is (EHOR, SPHOR(1)).

In the next stage we applied a discriminant analysis algorithm using the following discriminant function d_i for the ith class:

$$d_i(\boldsymbol{x}) = 2p_i - \log \det(S_i) - (\boldsymbol{x}-\boldsymbol{m}_i)^{\mathrm{T}} S_i^{-1}(\boldsymbol{x}-\boldsymbol{m}_i)$$

where $\boldsymbol{x}$ is the vector of the parameters of the object being examined, $\boldsymbol{m}_i$ is the mean vector of the ith class and S_i is the sample covariance matrix of the ith class. p_i is the *a priori* probability that an object belongs to the ith class and is proportional to the number of times that it is observed.

Table 3 *Results of the discriminant analysis*

Class	*Number of observations*	*Classification for vertical/horizontal case*		
		1	*2*	*3*
1	42	39/41	3/1	0/0
2	67	4/2	55/56	8/9
3	65	1/0	2/4	62/61

The classification was carried out for two variants: (i) the 16 vertical parameters; (ii) the 22 horizontal parameters. The results are summarized in Table 3. It can be seen that the discriminant analysis algorithm correctly identifies the class of an accelerogram in 90% of cases for (i) and in 91% of cases for (ii). The result is much better than that obtained in our previous study of the dynamic parameters of strong tremor records. This may be due to the inclusion of spectral parameters in the present analysis.

It is useful to observe how the main parameters selected by the statistical algorithms are distributed with respect to the intensity values. When the intensity is plotted against $\log_{10}$SPVERT(4) (Fig. 1) we find that the threshold separating records with $I \leqslant 6$ from those with $I \geqslant 7$ is at $\log_{10}$SPVERT(4) = 2.8. There are 22 errors (two accelerograms with $I \geqslant 7$ have $\log_{10}$SPVERT(4) < 2.8 and 20 accelerograms with $I \leqslant 6$ have $\log_{10}$SPVERT(4) > 2.8). Analogously, only six accelerograms in the second class and no accelerograms in the first class have $\log_{10}$SPHOR(1) > 3.4, and 20 accelerograms in the third class have $\log_{10}$SPHOR(1) < 3.4 (Fig. 2). Thus if an accelerogram has $\log_{10}$SPHOR(1) > 3.4, there is a 90% probability that $I \geqslant 7$ and a 10% probability that $I = 6$. However, if $\log_{10}$SPHOR(1) < 3.4, there is an 84% probability that $I \leqslant 6$ and a 16% probability that $I \geqslant 7$. Thus we can easily distinguish destructive and moderate accelerograms.

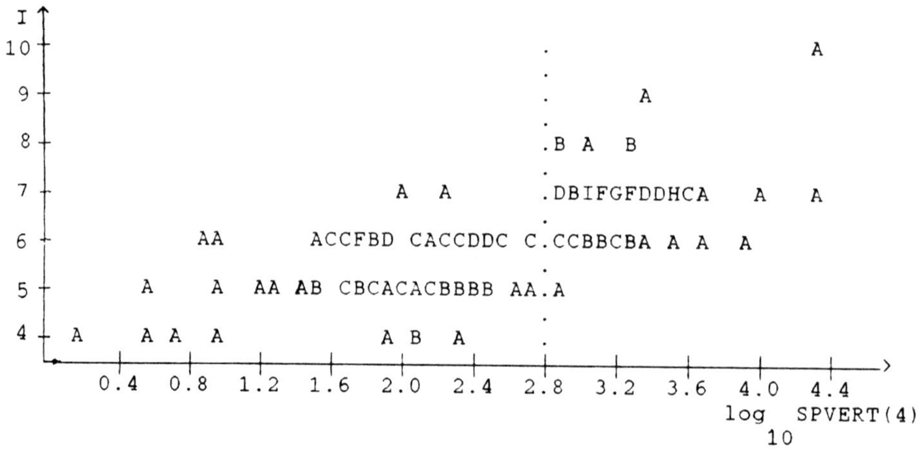

Fig. 1 *Plot of I versus $\log_{10}$SPVERT(4): A, one object; B, two objects etc.*

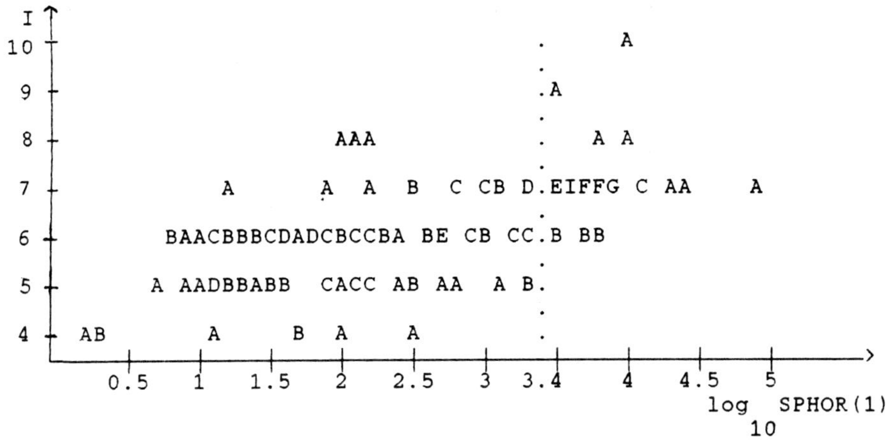

Fig. 2 *Plot of I versus* $\log_{10}$*SPHOR(1): A, one object; B, two objects etc.*

Again we see that there is a marked difference between the low frequency parameters of accelerograms from the destructive class and those from the other classes. These differences are particularly obvious in the interval 0.25–2.5 Hz. At higher frequencies the distribution of the parameters of various classes is more uniform.

4 Characteristic features of slight and destructive accelerograms

The purpose of the pattern-recognition algorithm Cora-3 is the detection of characteristic features of objects in two given classes (for a description of the algorithm see Cisternas *et al.* (1985)). Accordingly we consider here only records in the first and third classes and attempt to find which combinations of parameters can be used to distinguish slight and destructive accelerograms. The following result is obtained.

For a record to belong to the destructive class it is necessary and sufficient for it to possess one of the following two properties:

$$\text{SPVERT}(1) > 100 \text{ and } \text{SPVERT}(4) > 600$$
$$\text{SPHOR}(4) > 2500 \text{ and } \text{SPVERT}(4) < 250$$

Otherwise, $I < 6$. The first property is the most important because it is possessed by 61 out of 65 accelerograms with $I > 7$.

Cora-3 selects only the spectral parameters SPVERT(4), SPHOR(4) and SPVERT(1) for its decision rule which correctly distinguishes accelerograms of given strong tremors of the first and the third classes. As in the paper by Gvishiani *et al.* (1987) all objects from the first and third classes were included in the learning classes, and thus we examined all the learning objects. However, in the paper by Gvishiani *et al.* (1986) only 20 objects from each class were selected as learning material.

5 Conclusion

In this paper in which we continue our studies of the macroseismic significance of various parameters of strong tremors (Gvishiani *et al.*, 1986, 1987) we show that the values of the spectral energies at low frequencies (0.25–2.5 Hz) should be included in the major characteristics of accelerograms. In our earlier work we considered the peak horizontal velocity (Gvishiani *et al.*, 1986) and the "effective" acceleration (Gvishiani *et al.*, 1987) and duration (Schenk, 1985) as the most important parameters. The results of the statistical classification obtained in this work are much better than those obtained by Gvishiani *et al.* (1987) and show a direct relationship between the energy levels of the spectrum at low frequencies and the macroseismic effects of the oscillations. The methods used in this investigation are similar to those used by Boissonnade *et al.* (1984) and may help to solve the problem of determining macroseismic intensities by means of instruments.

References

BOISSONNADE A. C., DONG W. M., LIU S. C. and SHAH H. C. (1984): Use of pattern recognition and Bayesian classification for earthquake intensity and damage estimation, *Soil Dynamics and Earthquake Engineering*, **3**, 145–149.

CALIFORNIA INSTITUTE OF TECHNOLOGY (1969–1976): *Strong-motion Earthquake Accelerograms*, California Institute of Technology, Pasadena, CA.

CISTERNAS A., GODEFROY P., GVISHIANI A. D. *et al.* (1985): A dual approach to recognition of earthquake-prone areas in the Western Alps, *Annales de Geophysique*, **3**, 249–270.

ENSLEIN K., RALSTON A. and WILF H. (eds) (1977): *Statistical Methods for Digital Computers*, Wiley, New York.

GVISHIANI A. D., MOSTINSKY A. Z., TUMARKIN A. G. and ZHIZHIN M. N. (1986): Classification of strong motions by pattern recognition algorithms. In *Mathematical Methods of Analysis of Geophysical Information*, pp. 136–155, Academy of Sciences of the USSR, Moscow.

GVISHIANI A. D., SCHENK V. *et al.* (1987): Classification of strong ground motions according to macroseismic intensity by pattern recognition algorithms. In MAYER-ROSA D., VAN GILLS J. M. and STILLER H. (eds), *Proceedings of the 20th General Assembly of the ESC, Zurich.*

SCHENK V. (1985): Amplitude–frequency distribution of strong ground motion, *Physics of the Earth and Planetary Interiors*, **38**, 144–161.

Seismic Response and Stability of Earth Dams

S. K. SARMA

Civil Engineering Department, Imperial College, London SW7 2BU, Great Britain

Abstract The response of earth dams to earthquakes depends on many factors, one of which is the loss of energy through the foundations. In this paper a radiating boundary condition is provided which is also applied to an earth dam subjected to a sine pulse. The non-linear soil properties are made equivalent to linear properties with viscous damping. The yield strength of the soil depends on the rise of pore-water pressure during earthquake loading. The yield strength is characterized by the critical acceleration parameter which is determined by the limit equilibrium method of slope stability analysis which utilizes cyclic pore-pressure parameters. If the response shows that the critical acceleration is less than the maximum earthquake acceleration applied on a possible slipped mass, the displacements of the slipped mass can be computed using a simple formula.

1 Response of the structure

When subjected to earthquakes earth dams will respond in a manner which is dictated by (a) the properties of the material of the dam body and their capacity to absorb energy, (b) the properties of the foundation material, (c) the geometry of the structure itself and (d) the characteristics of the earthquake. Depending on these factors, we have to formulate a mathematical model, the solution to which gives the response of the structure to the earthquake. Many solutions to this model are available. For example, Ambraseys (1960), Seed and Martin (1966), Ambraseys and Sarma (1967), Sarma (1979b), Sarma (1980) and Gazetas (1980, 1982) have used one-dimensional elastic shear beam analysis, and Clough and Chopra (1966) and many others have used the finite element method.

Soil is an extremely non-linear material, the properties of which depend on factors such as the confining pressure, the void ratio, the degree of saturation etc. and above all on the stress path of loading. Since the future earthquake loading pattern is unknown, the mathematical model should be designed to be almost independent of the stress path. A convenient parameter for soil in this respect is the effective strength parameter. However, the actual strength will depend on the pore-water pressure which is controlled by the stress path. Since larger computers are now available, it is not difficult to introduce the non-linear behaviour of the material in an appropriate stress–strain relationship which includes the energy-absorbing capacity in the loading–unloading path. It is also possible to introduce the strength-degrading property of soils due to the increase in pore-water pressure or other factors (Zienkiewicz *et al.*, 1981; Prevost *et al.*, 1985). However, in terms of computing time or costs, these methods are very costly if the material properties and the time history of the earthquake are only approximately known. Therefore the use of such costly and time-consuming methods does not seem appropriate for design purposes. However, these methods are important for research purposes because they help us to calibrate simplified methods and to judge their usefulness. It is therefore reasonable to study the response of the dam by treating the material as elastic and remembering that this assumption is only a simplification and not reality.

Once we have decided that elastic response is acceptable, it is necessary to find the representative elastic properties. The most convenient elastic property is the shear wave velocity in the material which is given by the equation

$$S = \left(\frac{G}{\rho}\right)^{1/2} \tag{1}$$

where S is the shear wave velocity, G is the shear modulus and ρ is the mass density. Since the material is non-linear, the shear modulus (secant modulus) is strain dependent as shown in Fig. 1. Therefore we have to define a strain level, and the yield strain appears to be appropriate for this.

A second elastic parameter which may become important is Poisson's ratio ν. This parameter can be introduced directly or in the form of the compression wave velocity, which is defined as

$$C = S\left\{\frac{2(1-\nu)}{1-2\nu}\right\}^{1/2} \tag{2}$$

Furthermore, since elastic material does not lose energy, we have to introduce a parameter which represents the energy loss capacity. A convenient parameter in this respect is viscous damping.

However, internal friction (non-linearity) in the dam body is not the only mechanism by which the energy of the earthquake is lost. A major source of energy loss is through the foundation because it is not absolutely rigid.

We can account for this energy loss also by increasing the damping factor. However, there is a difference. While viscous damping changes the vibrational

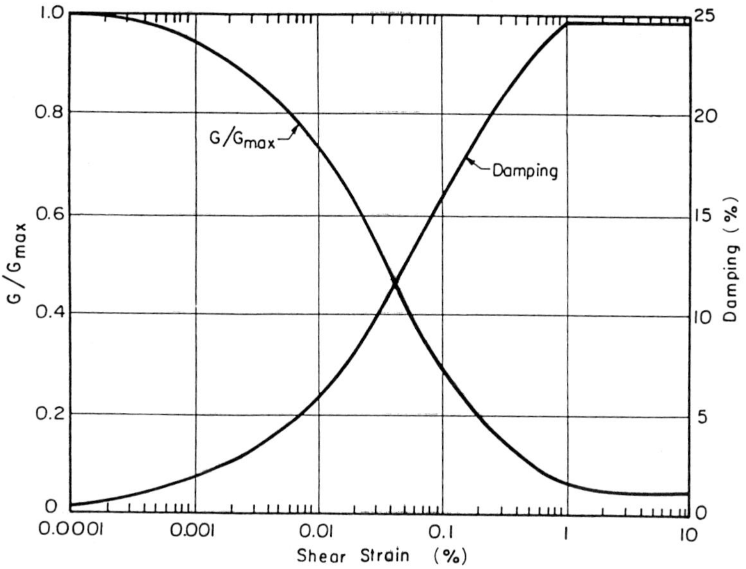

Fig. 1 *Shear modulus and damping characteristics used in the response computations (after Makdisi and Seed, 1977).*

characteristics of the structure, e.g. the mode frequency, the radiation energy loss through the foundation does not affect this characteristic. In the past, this energy loss was accounted for by putting the structure on viscous dampers at the boundary (Lysmer and Kuhlemeyer, 1969) or by introducing infinite elements at the boundary (Bettess, 1977; Medina and Penzien, 1982) or by using consistent boundary conditions (Lysmer and Wass, 1972). While viscous dampers can be used in time-domain analysis, the infinite element or consistent boundary methods can only be used in frequency-domain analysis. However, viscous dampers are correct only when one-dimensional analysis is applicable. For two-dimensional analysis, the choice of the boundary viscous damping parameters does not have a proper theoretical backing. Therefore we have undertaken a theoretical study of the radiation boundary condition. It appears that the boundary condition given below applies when the structure rests on the surface of another elastic material. This boundary condition is derived in two dimensions for the plane strain regime. However, extension to three dimensions is not difficult. It is also believed that the boundary condition can be modified for subsurface conditions, e.g. when the rock base is overlain by a superficial deposit. We have not as yet looked into the formulation of this part of the problem. The condition can be expressed as follows:

$$qC_f\left(\frac{\partial u}{\partial x}+\frac{\partial v}{\partial y}\right)+rS_s\left(\frac{\partial u}{\partial y}+\frac{\partial v}{\partial x}\right)=\dot{u}+\dot{v}-\dot{u}_g-\dot{v}_g \tag{3}$$

The subscripts s and f refer to the structure and the foundation rock respectively. C is

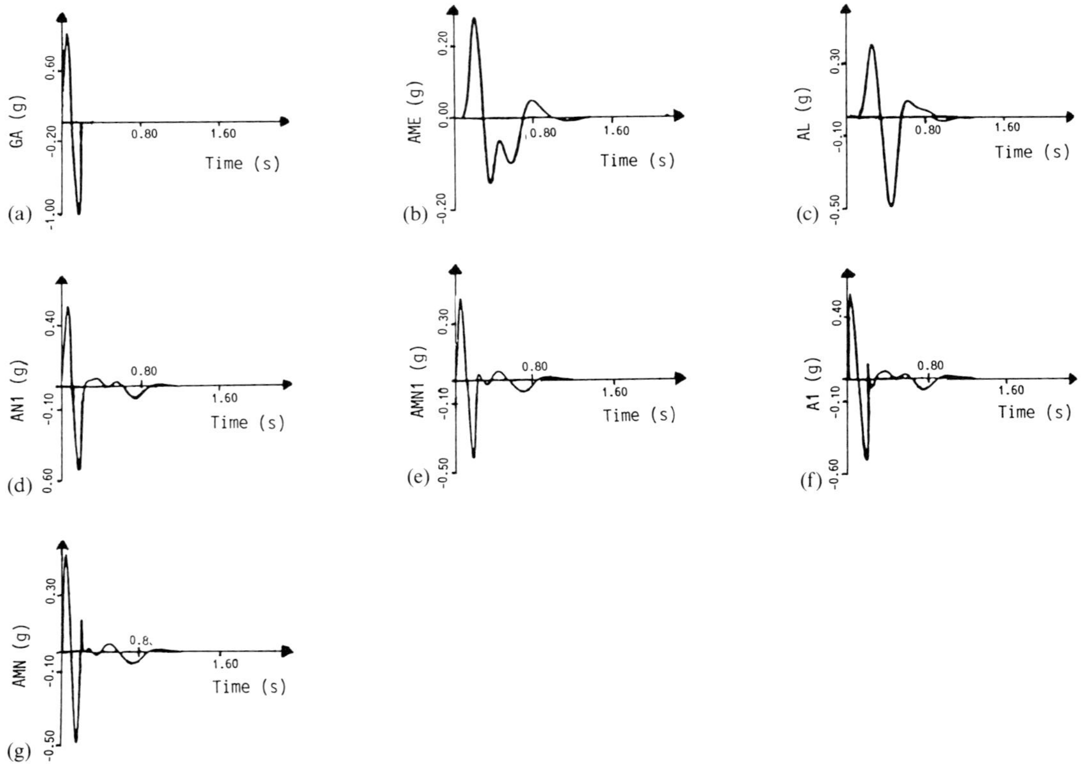

Fig. 2 *Acceleration time history for an earth dam: (a) free-field record; (b) centre of the dam; (c) crest of the dam; (d) edge of the dam just above the base; (e) middle of the dam just above the base; (f) edge of the dam at the foundation level; (g) middle of the dam at the foundation level; height $H = 60\,m$; base width $B = 195\,m$; crest width $B_1 = 15\,m$; number L of lumped mass, 61; Young's modulus $E = 195\,500\,kN\,m^{-2}$; $\nu = 0.43$; $\rho = 1.7\,kN\,m^{-3}$; equivalent viscous damping parameter $\lambda = 5.0\%$; $r = 1.0$; earthquake record EQK = sin 6; total duration $T = 0.2\,s$; time interval ΔT between points, 0.01 s.*

the compression wave velocity, S is the shear wave velocity,

$$q = \frac{C_s^2 - S_s^2}{C_f^2 - S_f^2} \frac{\rho_s}{\rho_f}$$

$$r = \frac{\rho_s S_s}{\rho_f S_f}$$

$\dot{u}_g(t)$ is the free-field ground velocity in the horizontal direction, $\dot{v}_g(t)$ is the free-field ground velocity in the vertical direction, $u(x, y, t)$ is the absolute horizontal displacement in the body of the structure and $v(x, y, t)$ is the absolute vertical displacement in the body of the structure. Unlike the boundary viscous dampers, both the horizontal and vertical motions are interlinked with the compression and shear wave velocities at the boundary, and in a one-dimensional analysis involving shear only, i.e. $q = 0$ and $v = 0$, it is equivalent to the viscous damper. We have introduced this boundary condition in a lumped-mass model solution without difficulty. This part of the research is still at the development stage. Typical results for the analysis with harmonic input are shown in Fig. 2 for various combinations.

2 The concept of critical acceleration

The response analysis gives the inertia forces that the structure is required to withstand, provided of course that the materials remain elastic. As has been noted above, soil is an extremely non-linear material which has a yield strength, and therefore the stresses implied by the elastic response analysis may be in excess of the yield strength in parts of the dam body. In that case the elastic response analysis is theoretically no longer valid. However, a complete non-linear analysis is also unnecessary in the context of poorly known material properties, dynamic pore-water pressure conditions and unknown future earthquake conditions. Therefore we approach the problem of the yield strength of the soil in a simplified way. The yield strength of the soil can be characterized by the critical acceleration which is defined as that horizontal acceleration which is needed to bring the soil strength to its yield limit. However, in the embankment and slope problem we look for the critical acceleration, not at some specific point, but for a mass of soil which may actually fail along a slip surface (Fig. 3). To find the critical acceleration, we apply the limit equilibrium analysis technique and we can modify any existing static stability analysis method for this purpose. However, two methods in particular (Sarma, 1973, 1979a) are useful for finding the critical acceleration of a possible slip surface. In finding the critical acceleration, we assume that the acceleration is an average value and is constant for the particular sliding mass. The response analysis shows that the acceleration at different points of the dam or the embankment is usually different at any instant in time. It is usually higher towards the top of the dam. However, the use of average and constant acceleration for a given sliding mass is conservative, and therefore it can be employed without hesitation.

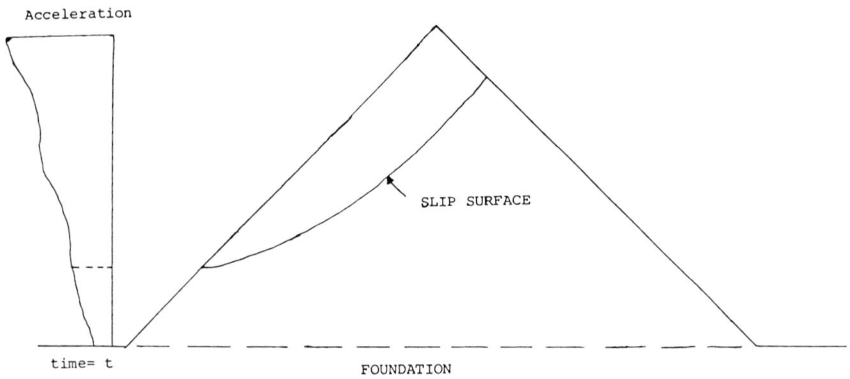

Fig. 3

Any soil strength parameters, such as effective strength parameters with undrained pore pressures or total strength parameters, can be used in the limit equilibrium analysis technique to find the critical acceleration. The use of effective strength with pore-water pressure is recommended. We have to remember that pore-water pressures may increase during the cyclic loading and unloading process. We can introduce the dynamic pore-water pressure generation in the limit equilibrium analysis by using a modified Skempton A parameter.

3 Dynamic pore-water pressure

The excess pore-water pressure generated during dynamic loading can be expressed by the relationships (Sarma and Jennings, 1980) (Fig. 4)

$$\frac{\Delta u}{\sigma_{3c}'} = A_n \frac{\Delta \sigma_1}{\sigma_{3c}'} \tag{4}$$

for the triaxial loading condition and

$$\frac{\Delta u}{\sigma_{v0}'} = A_n \frac{\Delta \tau}{\sigma_{v0}'} \tag{5}$$

for the simple shear condition. The parameter A_n is given by

$$A_n^{1/2} = A_1^{1/2} + \beta \log n \tag{6}$$

where A_1 is the parameter for one cycle loading, β is a constant and n is the number of cycles. The A_n parameters are different for different stress paths.

Equations (4) and (5) suggest that the Skempton (1954) equation can also be adopted for dynamic pore pressure and we can write

$$\Delta u = B\{\Delta \sigma_3 + A_n(\Delta \sigma_1 - \Delta \sigma_3)\} \tag{7}$$

We believe that the pore-water pressure rise during earthquake loading in an

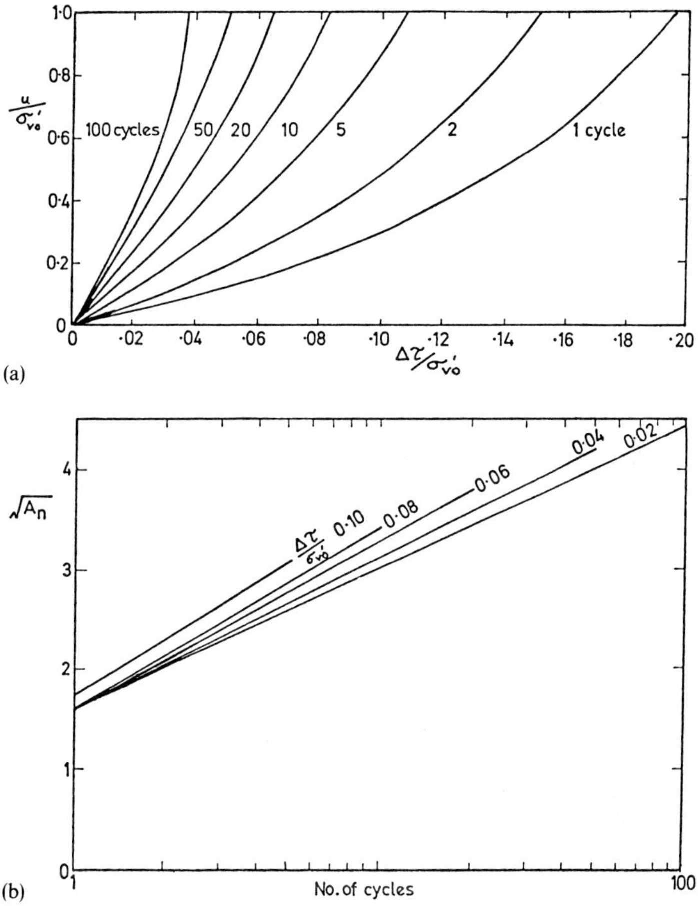

Fig. 4 *(a) Pore-pressure response in a simple shear test based on a numerical model for quartz sand (relative density, 45%; cyclic simple shear, $\sigma_{vo}' = 192\,kPa$); (b) pore-pressure parameters under cyclic loading ($A_n^{1/2} = A_1^{1/2} + \beta \log n$ where $A_1 = 2.6$, $\beta = 1.5$). (Data from Martin et al., 1975.)*

embankment can be described by an expression which is a combination of Eqns (7) and (5), but unfortunately we do not have any parameters to justify this statement.

4 Analysis

We describe below a modified version of Sarma's (1973) method of analysis which includes the dynamic pore-water pressure rise under triaxial conditions. The modification to simple shear conditions or combined conditions is not difficult. This method was originally published in a report to the US Corps of Engineers (Sarma,

1979b). It does not deal with liquefaction as such, but with stresses and pore pressures up to the initial failure stage. What happens beyond this stage is dealt with by the method of sliding-block movements (Newmark, 1965; Sarma, 1975).

A two-stage approach is adopted in the analysis. First we find the possible static stresses by performing a static analysis and then we introduce the earthquake load with dynamic pore-pressure parameters. This is a modified slice method.

Let us consider a possible slip surface in an embankment or a dam as shown in Fig. 5. We subdivide the sliding body into n vertical slices. Before the earthquake, slice i is acted upon by forces as shown in the figure. We denote pre-earthquake conditions by the subscript 0 and the earthquake conditions by the subscript d. The subscript i denoting the slice number is omitted throughout. Let us assume that we have solved the static stability problem and found the safety factor F_0, the corresponding interslice body forces and the normal and shear forces at the base of the slice by using any static stability analysis method (e.g. Sarma, 1973).

Let us now assume that a horizontal earthquake load kW is applied, under the influence of which limiting equilibrium is reached. In this case k is equal to the critical acceleration factor k_c. We can write the vertical and horizontal equilibrium equations of the slice as follows:

$$N_d \cos\alpha + T_d \sin\alpha = W - \delta X_d \tag{8}$$

and

$$T_d \cos\alpha - N_d \sin\alpha = k_c W + \delta E_d \tag{9}$$

Because the slice is in the limiting equilibrium condition, the Mohr–Coulomb failure criterion gives

$$T_d = (N_d - U_0 - \delta U_d)\tan\phi' + c'b\sec\alpha \tag{10}$$

where U_0 is the force due to static pore-pressure conditions and δU_d is the force due to the excess pore pressure on the base of the slice under the influence of $k_c W$. Since k_c is unknown at this stage, δU_d is also unknown.

At this point, we introduce the hypothesis due to Sarma (1975). This hypothesis states that if a possible failure plane is acted upon by a normal stress and a shear stress and the shear stress has a safety factor F, the state of stress in all planes will be the same as if the material has a strength property given by $\psi' = \tan^{-1}(\tan\phi'/F)$ and $\bar{c}' = c'/F$. According to this hypothesis, it is possible to draw Mohr's stress circle for static conditions with a safety factor F_0, as shown in Fig. 5(b). From the geometry of the circle, the major and minor effective stresses before the earthquake are given by

$$\sigma_{1,0}' = \sigma_0' + \tau_0(\tan\psi_0' + \sec\psi_0') \tag{11}$$

$$\sigma_{3,0}' = \sigma_0' + \tau_0(\tan\psi_0' - \sec\psi_0') \tag{12}$$

where

$$c_0' = \frac{c'}{F_0}$$

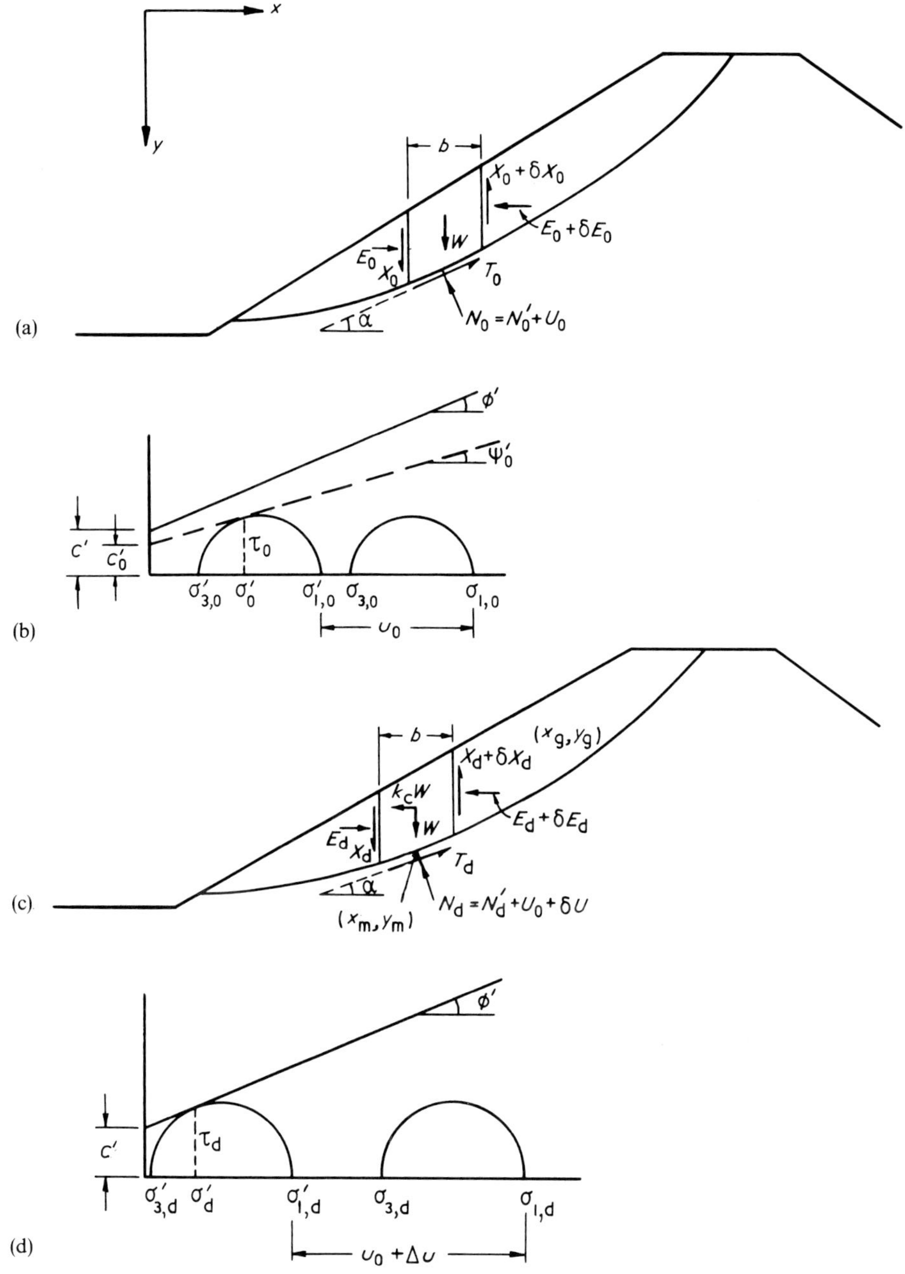

Fig. 5 *Stability analysis by the Sarma method: (b) static condition, safety factor* F_0; $c_0' = c'/F_0$; $\psi_0' = \tan^{-1}(\tan\phi'/F_0)$; $\sigma_0' = N_0'/(b\sec\alpha)$; $\tau_0 = T_0/(b\sec\alpha)$; $u_0 = U_0/(b\sec\alpha)$; *(d) during an earthquake, critical acceleration* k_c; $\Delta U = \delta U/(b\sec\alpha)$; $\sigma_d' = N_d'/(b\sec\alpha)$; $\tau_d = T_d/(b\sec\alpha)$.

$$\psi_0' = \tan^{-1}\frac{\tan\phi'}{F_0}$$

When the static pore pressure u_0 at the base of the slice is added to both sides, the total stresses are given by

$$\sigma_{1,0} = \sigma_0 + \tau_0(\tan\psi_0' + \sec\psi_0') \tag{13}$$

$$\sigma_{3,0} = \sigma_0 + \tau_0(\tan\psi_0' - \sec\psi_0') \tag{14}$$

Similarly, for dynamic loading conditions we have (Fig. 5(d))

$$\sigma_{1,\mathrm{d}} = \sigma_\mathrm{d} + \tau_\mathrm{d}(\tan\psi_\mathrm{d}' + \sec\psi_\mathrm{d}') \tag{15}$$

$$\sigma_{3,\mathrm{d}} = \sigma_\mathrm{d} + \tau_\mathrm{d}(\tan\psi_\mathrm{d}' - \sec\psi_\mathrm{d}') \tag{16}$$

We define

$$\Delta\sigma_1 = \sigma_{1,\mathrm{d}} - \sigma_{1,0} \tag{17}$$

and

$$\Delta\sigma_3 = \sigma_{3,\mathrm{d}} - \sigma_{3,0} \tag{18}$$

and therefore Eqn (7) gives

$$\Delta u = B\{\Delta\sigma_3 + A_n(\Delta\sigma_1 - \Delta\sigma_3)\}$$

Multiplying both sides by $b \sec\alpha$, we obtain for the critical earthquake loading condition with $\psi_\mathrm{d}' = \phi'$

$$\begin{aligned}\delta U_\mathrm{d} = B[(N_\mathrm{d} - N_0) + T_\mathrm{d}\{\tan\phi' - \sec\phi'(1-2A_n)\}\\ - T_0\{\tan\psi_0 - \sec\psi_0(1-2A_n)\}]\end{aligned} \tag{19}$$

Eliminating N_d and T_d from Eqn (19) by using Eqns (8) and (10), we obtain

$$\delta U_\mathrm{d} = \bar{a} - \bar{b}\delta X_\mathrm{d} \tag{20}$$

where

$$\bar{a} = \frac{B\{(a - N_0) + dq - T_0 q_0\}}{1 - B(b_0 - eq)} \tag{21}$$

$$\bar{b} = \frac{B(c + fq)}{1 - B(b_0 - eq)} \tag{22}$$

$$a = (W + U_0 \sin\alpha\tan\phi' - c'b\tan\alpha)\cos\phi'\sec(\phi' - \alpha) \tag{23}$$

$$b_0 = \frac{\tan\alpha\tan\phi'}{1 + \tan\alpha\tan\phi'} = e\tan\alpha \tag{24}$$

$$C = \cos\phi'\sec(\phi' - \alpha) \tag{25}$$

$$d = (c'b + W\tan\phi' - U_0\cos\alpha\tan\phi')\cos\phi'\sec(\phi' - \alpha) \tag{26}$$

$$e = \frac{\tan \phi'}{1 + \tan \alpha \tan \phi'} \tag{27}$$

$$f = \sin \phi' \sec(\phi' - \alpha) \tag{28}$$

$$q = \tan \phi' - \sec \phi' (1 - 2A_n) \tag{29}$$

$$q_0 = \tan \psi_0' - \sec \psi_0' (1 - 2A_n) \tag{30}$$

$$N_0 = (W + U_0 \sin \alpha \tan \psi_0' - c_0' b \tan \alpha) \cos \psi_0' \sec(\psi_0 - \alpha) - \delta X_0 \cos \psi_0' \sec(\psi_0' - \alpha) \tag{31}$$

$$T_0 = (c_0' b + W \tan \psi_0' - U_0 \cos \alpha \tan \psi_0') \cos \psi_0' \sec(\psi_0' - \alpha) - \delta X_0 \sin \psi_0' \sec(\psi_0' - \alpha) \tag{32}$$

$$\psi_0' = \tan^{-1}\left(\frac{\tan \phi'}{F_0}\right) \tag{33}$$

$$c_0' = \frac{c'}{F_0} \tag{34}$$

Now eliminating N_d and T_d from Eqns (8), (9) and (10), we obtain

$$\delta X_d \tan(\phi' - \alpha) + \delta E_d = W \tan(\phi' - \alpha) + (c' b \sec \alpha \cos \phi' - U_0 \sin \phi') \sec(\phi' - \alpha) - (\bar{a} - \bar{b} \delta X_d) \sin \phi' \sec(\phi' - \alpha) - k_c W$$

Rearranging terms gives

$$\delta X_d \{\tan(\phi' - \alpha) - \bar{b} \sin \phi' \sec(\phi' - \alpha)\} + \delta E_d = \bar{D} - k_c W \tag{35}$$

where

$$\bar{D} = W \tan(\phi' - \alpha) + (c' b \sec \alpha \cos \phi' - U_0 \sin \phi') \sec(\phi' - \alpha) - \bar{a} \sin \phi' \sec(\phi' - \alpha) \tag{36}$$

Since in the absence of other external forces $\Sigma \delta E_d = 0$, summing for all slices gives

$$\Sigma \delta X_d \{\tan(\phi' - \alpha) - \bar{b} \sin \phi' \sec(\phi' - \alpha)\} + k_c \Sigma W = \Sigma \bar{D} \tag{37}$$

Let us assume that the point of application of the forces N_d is known and that the coordinate of this point is (x_m, y_m) which, for practical purposes, can be taken as the midpoint of the base of the slice. Then, taking moments about the centre of gravity (x_g, y_g) of the whole sliding body and remembering that the net moment of W and $k_c W$ about the centre of gravity is zero, and with the reference axes as shown in Fig. 5, we have

$$\Sigma(N_d \cos \alpha + T_d \sin \alpha)(x_m - x_g) + \Sigma(T_d \cos \alpha - N_d \sin \alpha)(y_m - y_g) = 0 \tag{38}$$

Using Eqns (8), (9) and (35) we obtain

$$\Sigma \delta X_d [(y_m - y_g)\{\tan(\phi' - \alpha) - \bar{b} \sin \phi' \sec(\phi' - \alpha)\} + (x_m - x_g)] = \Sigma W(x_m - x_g) + \Sigma \bar{D}(y_m - y_g) \tag{39}$$

The problem now is to find a set of forces X_d which will satisfy Eqn (39) and at the same time will satisfy the condition that in the absence of other external forces $\Sigma\delta X_{\mathrm{d}} = 0$.

Let us assume that the shape of the X_{d} force distribution is known but the magnitude is not. Then we can say that

$$X_{\mathrm{d}} = \lambda Q \tag{40}$$

where Q is a known distribution but λ is unknown. Then, for the ith slice,

$$\begin{aligned} \delta X_{\mathrm{d}} &= \lambda(Q_{i+1} - Q_i) \\ &= \lambda P \end{aligned}$$

Equations (37) and (39) become

$$\lambda\Sigma P\{\tan(\phi' - \alpha) - \bar{b}\sin\phi'\sec(\phi' - \alpha)\} + k_{\mathrm{c}}\Sigma W = \Sigma\bar{D} \tag{41}$$

$$\begin{aligned} \lambda\Sigma P[(x_{\mathrm{m}} - x_{\mathrm{g}}) + (y_{\mathrm{m}} - y_{\mathrm{g}})\{\tan(\phi' - \alpha) - \bar{b}\sin\phi'\sec(\phi' - \alpha)\}] \\ = \Sigma W(x_{\mathrm{m}} - x_{\mathrm{g}}) + \Sigma\bar{D}(y_{\mathrm{m}} - y_{\mathrm{g}}) \end{aligned} \tag{42}$$

This gives

$$\lambda = S_4/S_3 \tag{43}$$

and

$$k_{\mathrm{c}} = \frac{S_1 - \lambda S_2}{\Sigma W} \tag{44}$$

where

$$\bar{D} = W\tan(\phi' - \alpha) + (c'b\sec\alpha\cos\phi' - U_0\sin\phi')\sec(\phi' - \alpha) - a\sin\phi'\sec(\phi' - \alpha) \tag{45}$$

$$S_1 = \Sigma\bar{D} \tag{46}$$

$$S_2 = \Sigma P\{\tan(\phi' - \alpha) - \bar{b}\sin\phi'\sec(\phi' - \alpha)\} \tag{47}$$

$$S_3 = \Sigma P[(y_{\mathrm{m}} - y_{\mathrm{g}})\{\tan(\phi' - \alpha) - \bar{b}\sin\phi'\sec(\phi' - \alpha)\} + (x_{\mathrm{m}} - x_{\mathrm{g}})] \tag{48}$$

$$S_4 = \Sigma W(x_{\mathrm{m}} - x_{\mathrm{g}}) + \Sigma\bar{D}(y_{\mathrm{m}} - y_{\mathrm{g}}) \tag{49}$$

Once λ and k_{c} have been determined, the interslice forces can be found from Eqns (40) and (35), and the normal and shear forces at the base of the slice can be derived. All these forces must be acceptable.

The function Q (Eqn (40)) is defined by Sarma (1973) as

$$Q = \mu\left\{(K' - R_{\mathrm{u}})\bar{\gamma}h^2\tan\left(\frac{\bar{\phi}'}{2}\right) + \bar{c}'h\right\}$$

where

$$K' = \frac{1 - \sin\beta\{(1 - 2\bar{R}_{\mathrm{u}})\sin\phi' + 4c'\cos\phi'/\gamma h\}}{1 + \sin\beta\sin\phi'} \tag{51}$$

$$\beta = 2\alpha - \phi' \tag{52}$$

$$\bar{R}_u = \frac{2P_w}{\gamma h^2} \tag{53}$$

μ is an arbitrary number, which is usually unity, P_w is the water pressure on the side of the slice, $\bar{\gamma}$ is the weighted average density for the vertical section, $\bar{\phi}'$ is the weighted average friction angle for the vertical section, $\bar{c}'$ is the weighted average cohesion for the vertical section, and ϕ' and c' are shear strength parameters for the base of the slice.

An alternative to using the parameter A_n is to estimate the pore pressures in advance of the earthquake load or to estimate the total strength. The use of the total strength is reasonable if we are trying to find the safety factor for a given earthquake load. However, the use of effective strength is preferable since the failure surface in the laboratory samples appears to be controlled by the effective strength parameters of the soil.

Seed and coworkers (Seed *et al.*, 1973; Seed, 1979) have used a very different approach, particularly in relation to liquefaction. Their method involves the following stages.

(1) Determine the possible dynamic stresses on the body of the dam using either a linear or a pseudo-non-linear approach. They advocate the use of the finite element method.

(2) On the basis of these dynamic stresses, find from cyclic laboratory tests whether liquefaction is possible in different parts of the dam.

(3) If liquefaction is likely to occur, check the stability of the dam after the earthquake, assuming zero shear strength in these zones.

(4) If no instability is predicted, then the dam is considered to be safe.

5 The concept of the displacement of the sliding mass

If the maximum applied inertia acceleration $k_m g$ is greater than the critical acceleration $k_c g$, the mass will slide along the slip surface. It should be emphasized that the inertia forces are computed on the basis that the material is elastic and therefore does not fail. However, the critical acceleration implies that the structure

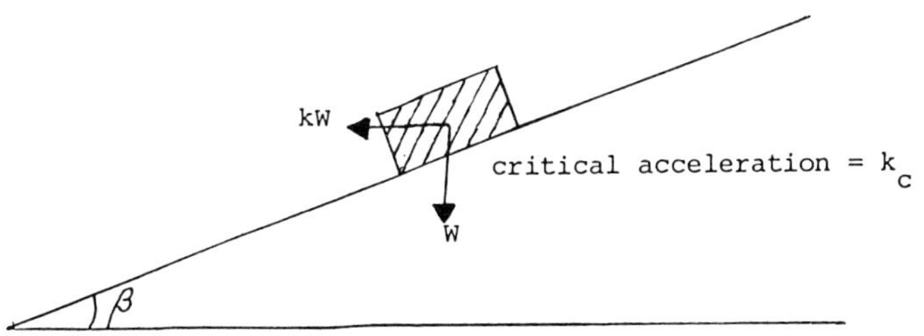

Fig. 6 Sliding-block model.

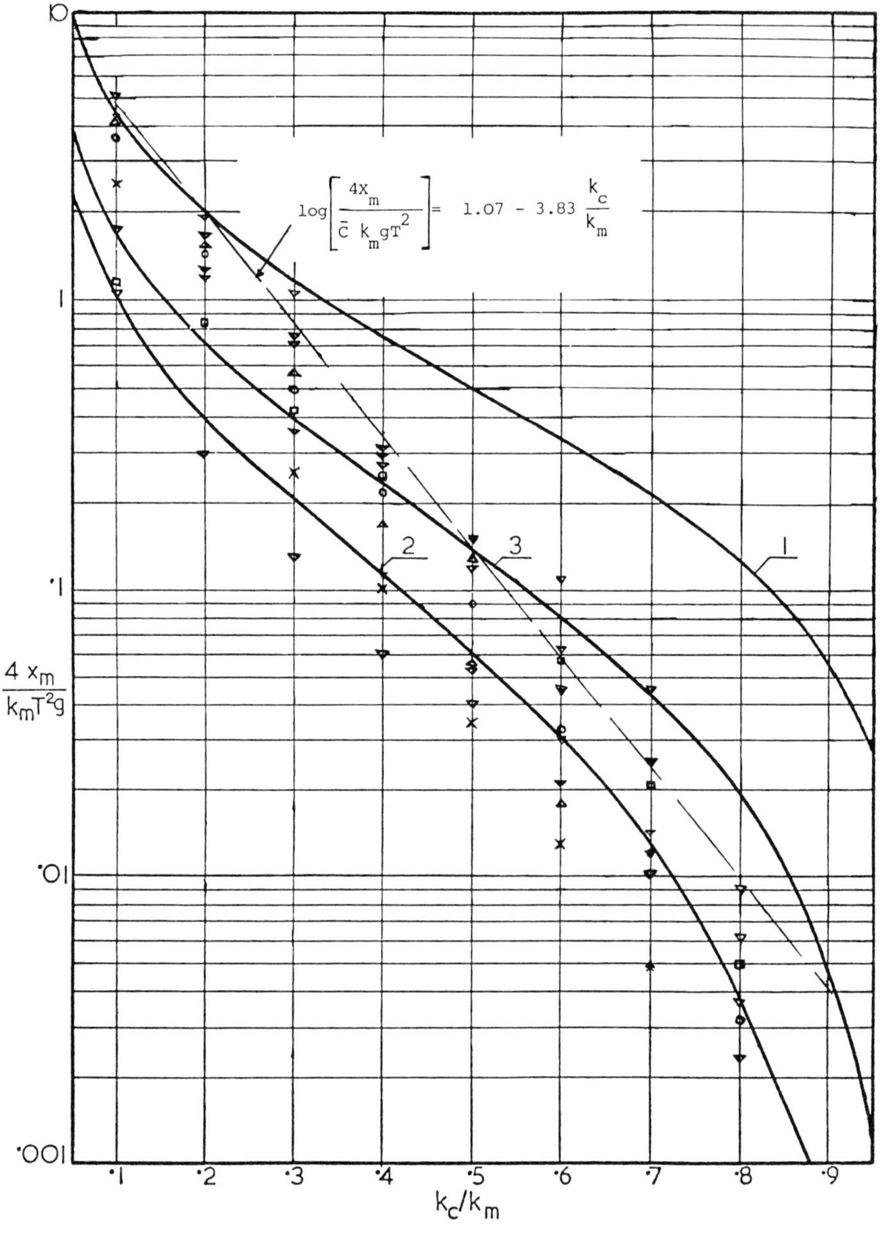

Fig. 7 *Plot of Eqn (54): 1, rectangular pulse; 2, triangular pulse; 3, half-sine pulse;* $\bar{C} = \{\cos(\theta+\beta-\phi')\}/\cos\ \phi'$; ○, *El Centro 1940, NS* ($T = 0.5$ s); △, *Koyna 1967, L* ($T = 0.15$ s); X, *Pacoima 1971, S16E* ($T = 0.4$ s); □, *Parkfield 2 1966, N65E* ($T = 0.6$ s); ▽, *several nuclear blasts.*

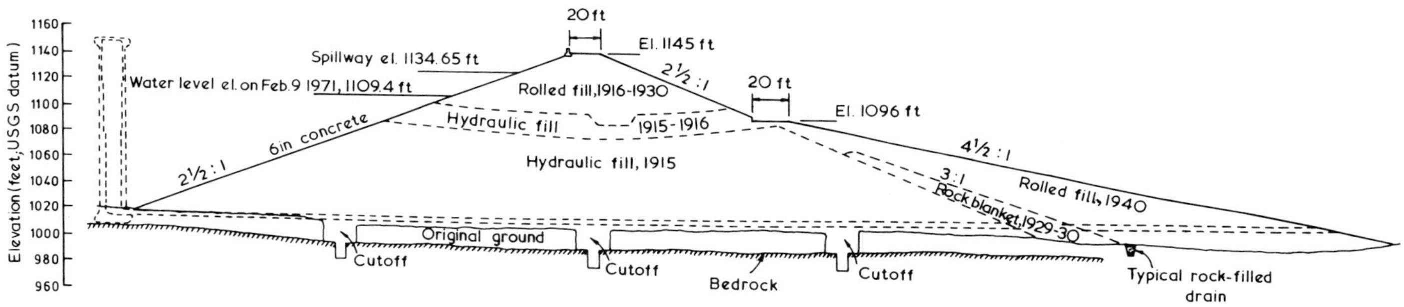

(a)

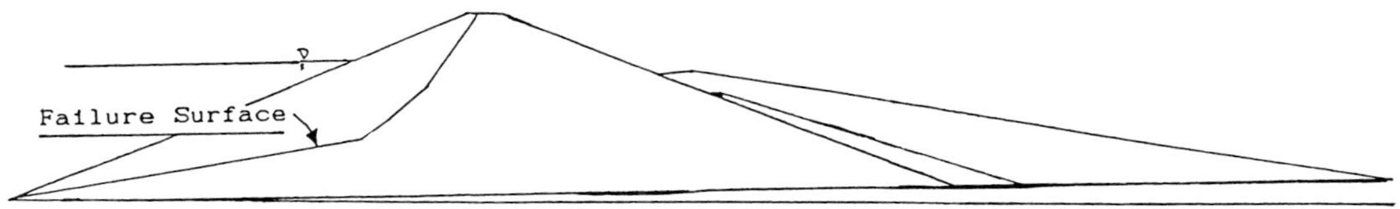

(b)

Fig. 8 *(a) Cross-section of the San Fernando Dam; (b) cross-section of the dam as used in the analysis showing the failure surface.*

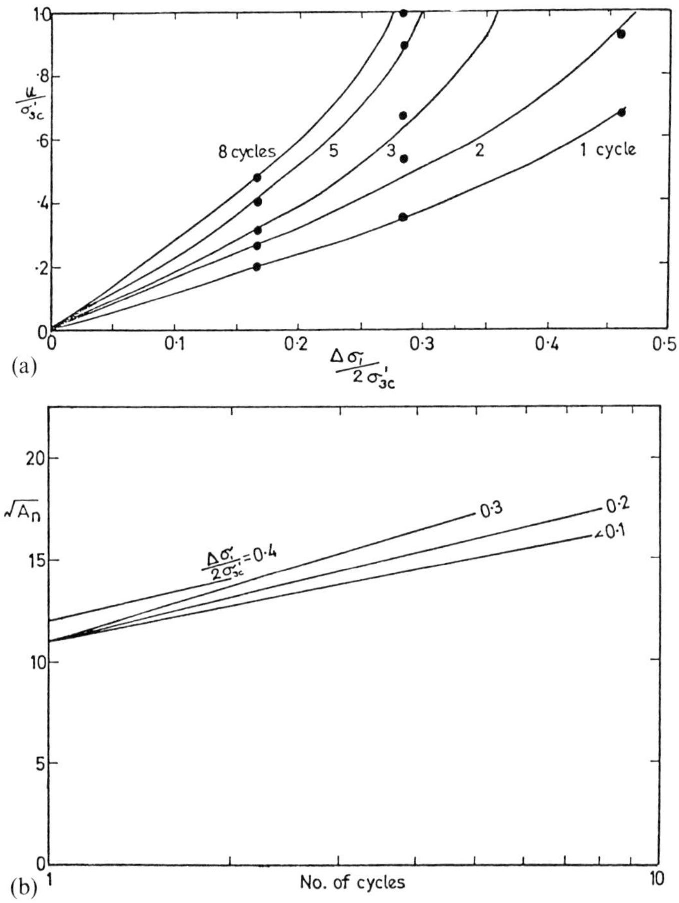

Fig. 9 *(a) Pore-pressure response in a cyclic triaxial test (●, data from Seed et al. (1973); hydraulic sand fill, Upper San Fernando Dam; $\sigma_{3c}' = 103\,kPa$); (b) pore-pressure parameters for San Fernando Dam material ($A_n^{1/2} = A_1^{1/2} + \beta \log n$ where $A_1 = 1.25$, $\beta = 0.7$) (note the difference in definition which results in a factor of 2 in the parameters compared with Fig. 4).*

actually fails along a slip surface after the stresses have reached a yield limit. The two statements are contradictory and the concept of the displacement of the sliding mass is introduced to bridge this contradiction (Newmark, 1965; Sarma, 1975). The concept is easy to understand and can be visualized using a sliding-block model (Fig. 6). The displacement of the sliding block can be computed using the following equation which was derived from the data given by Sarma (1975) and later tested using many other strong motion records (Fig. 7):

$$\frac{\log x_m}{\bar{C} k_m g T^2} = 1.07 - 3.83 \frac{k_c}{k_m} \tag{54}$$

where $\bar{C}$ is a constant with a value very close to unity.

6 Example

We take the Lower San Fernando Dam as an example for the study of critical acceleration. This dam failed in the San Fernando earthquake of 9 February 1971. The case has been very well studied (Seed *et al.*, 1973). The cross-section of the dam and the possible slip surface are shown in Fig. 8. Figure 9 shows the results of cyclic loading tests, which are replotted from the results reported by Seed *et al.* (1973). This gives $A_1 = 0.67$ and $A_5 = 1.35$. The results of the stability analysis are as follows.

If there is no dynamic pore-pressure rise $k_c = 0.21g$. This shows that there is a sufficient safety factor in the absence of an earthquake load. In a pseudo-static analysis with an acceleration of $0.21g$ the safety factor is equal to unity. During an earthquake the values shown in Table 1 are obtained. These values do not indicate liquefaction. The slope is statically safe after the earthquake since k_c is positive. Seed *et al.* (1973) assumed liquefied zones in parts of the dam and found a static safety factor of 1.06 which is comparable with the above results.

Table 1

A_n	k_c
0.4	0.16
0.8	0.13
1.8	0.11

7 Conclusion

The use of the dynamic pore-pressure parameter in the stability analysis makes it feasible to apply effective stress analysis to find the critical acceleration factor during earthquake loading. This factor can then be used in conjunction with a simplified response analysis to check the stability of the dam. If it is found to be unstable during the earthquake, sliding-block displacements can be computed using a simple relationship. The amount of the displacement will guide the designer as to the suitability of the structure. Although it is possible to analyse the response of structures with non-linear properties and a pore-water pressure rise, this approach should be used for research purposes and to calibrate the results of a simplified analysis such as that presented here.

References

Ambraseys N. N. (1960): The seismic stability of earth dams, *Proceedings of the 2nd World Conference on Earthquake Engineering*, Vol. 2, pp. 1345–1363, Science Council of Japan, Tokyo.

Ambraseys N. N. and Sarma S. K. (1967): The response of earth dams to strong earthquakes, *Geotechnique*, **17**, 181–213.

BETTESS P. (1977): Infinite elements, *International Journal for Numerical Methods in Engineering*, **2**, 53–64.

CLOUGH R. W. and CHOPRA A. K. (1966): Earthquake stress analysis in earth dams, *Journal of the Engineering Mechanics Division of the ASCE*, **92** (2), 197–211.

GAZETAS G. (1980): 3-D lateral and longitudinal seismic stability of earth and rock fill dams, *Proceedings of the 7th World Conference on Earthquake Engineering, Istanbul*, pp. 109–118, Turkish National Committee on Earthquake Engineering, Ankara.

GAZETAS G. (1982): Shear vibrations of vertically inhomogeneous earth dams, *International Journal for Numerical and Analytical Methods in Geomechanics*, **6** (1), 219–241.

LYSMER J. and KUHLEMEYER R. L. (1969): Finite dynamic model for infinite media, *Journal of the Engineering Mechanics Division of the ASCE*, **95** (4), 859–877.

LYSMER J. and WASS G. (1972): Shear waves in plane infinite structures, *Journal of the Engineering Mechanics Division of the ASCE*, **98** (1), 85–105.

MAKDISI F. I. and SEED H. B. (1977): A simplified procedure for estimating earthquake induced deformations in dams and embankments, *Report UCB/EERC-77/19*, University of California, Berkeley, CA.

MARTIN G. R., FINN W. D. L. and SEED H. B. (1975): Fundamentals of liquefaction under cyclic loading, *Journal of the Geotechnical Engineering Division of the ASCE*, **101** (5), 423–438.

MEDINA F. and PENZIEN J. (1982): Infinite elements for elastodynamics, *Earthquake Engineering and Structural Dynamics*, **10**, 699–709.

NEWMARK N. M. (1965): Effects of earthquakes on dams and embankments, *Geotechnique*, **15** (2), 139–160.

PREVOST J. H., ABDEL-GAFFER A. M. and LACY S. J. (1985): Non-linear dynamic analysis of an earth dam, *Journal of the Geotechnical Engineering Division of the ASCE*, **111** (7), 882–897.

SARMA S. K. (1973): Stability analysis of embankments and slopes, *Geotechnique*, **23** (3), 423–433.

SARMA S. K. (1975): Seismic stability of earth dams and embankments, *Geotechnique*, **25** (4), 743–761.

SARMA S. K. (1979a): Stability analysis of embankments and slopes, *Journal of the Geotechnical Engineering Division of the ASCE*, **105** (2), 1511–1524.

SARMA S. K. (1979b): Response and stability of earth dams during strong earthquakes, *Report Misc-Gl-79-13*, US Corps of Engineers, Vicksburg, MS.

SARMA S. K. (1980): A simplified method for the earthquake resistant design of earth dams, *Dams and Earthquakes, Proceedings of the ICE Conference, London*, pp. 155–160, Thomas Telford, London.

SARMA S. K. and JENNINGS D. N. (1980): A dynamic pore pressure parameter A_n. In PANDE G. N. and ZIENKIEWICZ O. C. (eds), *International Symposium on Soils under Cyclic and Transient Loading, Swansea*, pp. 295–298, Balkema, Rotterdam.

SEED H. B. (1979): Considerations in the earthquake resistant design of earth and rockfill dams, *Geotechnique*, **29** (3), 215–263.

SEED H. B. and MARTIN G. (1966): The seismic coefficient in earth dam design, *Journal of the Soil Mechanics and Foundation Division of the ASCE*, **92** (3), 25–58.

SEED H. B., LEE K. L., IDRISS I. M. and MAKDISI F. (1973): Analysis of the slides in the San Fernando Dams during the earthquake of February 9, 1971, *Report EERC-73-2*, University of California, Berkeley, CA.

SKEMPTON A. W. (1954): Pore pressure parameters *A* and *B*, *Geotechnique*, **4** (4), 143.

ZIENKIEWICZ O. C., LEUNG K. H., HINTON E. and CHANG C. T. (1981): Earth dam analysis for earthquakes: numerical solution and constitutive relations for non-linear (damage) analysis, *Dams and Earthquakes, Proceedings of the ICE Conference, London*, pp. 179–194, Thomas Telford, London.

Spectral Analysis of Seismic Ground Motion: Application to the Beznar Dam Accelerogram

J. SARRATE, J. A. CANAS, A. H. BARBAT AND J. EGOZCUE

Civil Engineering School, Polytechnic University of Catalonia, Barcelona, Spain

Abstract Numerical procedures for computing the Fourier amplitude spectrum and seismic response spectra are discussed. By making appropriate simple assumptions, it is shown that the velocity seismic response spectrum with zero damping can be approximated by the acceleration Fourier spectrum. A numerical comparison has been performed using a real seismic record from the Beznar Dam, Spain. The result suggests that this is a good approximation for high and intermediate frequency seismic excitation.

1 *Introduction*

Seismic ground motion records are generally obtained as real functions of time. The main objective of spectral analysis is to give a complete description of the frequency content of these records, which is used in seismological studies. The standard tool for obtaining a frequency content representation of seismic signals is Fourier analysis.

In earthquake engineering the frequency-domain representation of ground motion is used as a basic definition of seismic action. Typical frequency representations of seismic actions are the response spectra, which are suitable for application in the practical computation of the maximum response of structures. Thus the seismic response spectra represent standard frequency descriptions of seismic actions in earthquake engineering.

It is difficult to obtain the response spectra from the Fourier transform and vice versa, although efforts have been made to obtain practical relations (Housner, 1970; Duarte, 1987). Moreover this lack of intuitive relations leads to some misinterpretations of the Fourier spectrum. The modulus of the discrete Fourier transform of ground acceleration is used in two different ways:

(1) as a deterministic Fourier spectrum;

(2) as an intermediate step in the estimation of the spectral density of stationary stochastic processes which model the ground acceleration.

The first approach allows the ordinates of the Fourier spectrum to be interpreted as amplitudes of harmonic frequencies, and therefore they have acceleration units. In the second approach the areas under the spectral density plots are interpreted as variances of acceleration (squared acceleration units) associated with each frequency interval. These two options do not give any direct information about the energy introduced in a single-degree-of-freedom system by the input acceleration. However, it is possible to obtain a very simple relation between the input acceleration Fourier spectrum and the undamped velocity response spectrum. This relation has poor physical meaning but it is numerically suitable for engineering purposes.

The main objectives of this study are as follows:

(1) to put together several useful and important concepts that appear infrequently in the literature;

(2) to show analytically the relationship between the velocity response spectrum with zero damping and the Fourier amplitude spectrum;

(3) to apply the procedures described to the only accelerogram available, to our knowledge, for southern Spain.

The Fourier spectrum and response spectra, which are well-known concepts and computational tools, are summarized in Sections 2 and 3. In Section 4 it is shown theoretically and numerically how the undamped velocity response spectrum is an approximation of the acceleration Fourier spectrum.

2 Fourier amplitude spectrum

The Fourier transform of a general square integrable function $f(t)$ is defined as (Plancherel's theorem)

$$\begin{aligned} F(\theta) &= \int_{-\infty}^{\infty} f(t)\exp(-\mathrm{i}\theta t)\,\mathrm{d}t \\ f(t) &= \frac{1}{2\pi}\int_{-\infty}^{\infty} F(\theta)\exp(\mathrm{i}\theta t)\,\mathrm{d}\theta \end{aligned} \tag{1}$$

By taking the real and imaginary parts $\mathscr{R}\{F(\theta)\}$ and $\mathscr{I}\{F(\theta)\}$ of the Fourier transform (Eqn (1)) the Fourier amplitude spectrum is defined by

$$|F(\theta)| = ([\mathscr{R}\{F(\theta)\}]^2 + [\mathscr{I}\{F(\theta)\}]^2)^{1/2} \tag{2}$$

Since in seismology and earthquake engineering problems the signals are always finite, continuous and bounded, the integrals of the Fourier transform definitions (1) always exist and can be evaluated. Therefore the amplitude Fourier spectrum also exists and provides information about the amplitude of each frequency component.

In civil engineering applications the most important time function $f(t)$ is an accelerogram. However, in geophysical problems this function could be a displacement graph, a velocigram or an accelerogram. In any case the function is available in analogue or digital form. Thus, as it is not possible to use analytical methods to calculate the Fourier spectrum, numerical procedures have to be applied. The computation is carried out using the discrete Fourier transform

$$\tilde{F}(\theta) = \sum_{n=-\infty}^{\infty} f(n\Delta)\exp(\mathrm{i}\theta n\Delta) \qquad \theta_c = \frac{\pi}{\Delta} \qquad \theta \in [-\theta_c, \theta_c] \tag{3}$$

where the input signal $f(t)$ is sampled at regular intervals Δ. The seismic signals are finite, and thus $0 \leqslant t \leqslant T$. If $f(t)$ is sampled at N regular points, the limited discrete Fourier transform is defined as

$$F_N\left(k\frac{2\pi}{T}\right) = \sum_{j=0}^{N-1} f(j\Delta)\exp\left(-\mathrm{i}\frac{2\pi}{N}kj\right) \qquad k = 0, \ldots, N-1 \tag{4}$$

It should be noted that the value of $F_N(2\pi k/T)$ is equal to the value of F_N corresponding to the negative frequency $2\pi(k-N)/T$ when k is greater than $N/2$. Therefore only $N/2$ terms are needed to compute the amplitude Fourier spectrum. The Fourier transform (1) can be approximately computed through the limited discrete Fourier transform as follows:

$$F\left(k\frac{2\pi}{T}\right) \approx \Delta \sum_{j=0}^{N-1} f(j\Delta)\exp\left(-\mathrm{i}\frac{2\pi}{N}kj\right) = \Delta F_N\left(k\frac{2\pi}{T}\right) \qquad k = 0, \ldots, n \leqslant \frac{N}{2} \tag{5}$$

The sampling theorem (Tretter, 1976) shows that (5) is exactly true when $f(t)$ is a band-limited function in $[-\theta_c, \theta_c]$, i.e. $F(\theta) = 0$ if $|\theta| \geqslant \theta_c$. This method provides a good procedure for evaluating the Fourier transform of a general finite signal. There are many methods for calculating a limited discrete Fourier transform, but the most common is the fast Fourier transform. This has been discussed by many workers and extensive explanations have been given by Cooley *et al.* (1967), Singleton (1969), Brigham (1974), Tretter (1976) and many others. Computer programs are available in most computer libraries.

3 *Seismic response spectra and pseudo-response spectra*

3.1 Definitions and numerical computation

Consider a single-degree-of-freedom system whose motion is governed by the second-order differential equation

$$\ddot{x} + 2\nu\omega\dot{x} + \omega^2 x = -a(t) \tag{6}$$

where ω is the natural frequency of the system, ν is the damping ratio and $\omega_a = \omega(1-\nu^2)^{1/2}$ is the frequency of the damped system. The solution of this equation of motion is expressed in the form

$$x(t) = -\frac{1}{\omega_a}\int_0^t a(\tau)\exp\{-\nu\omega(t-\tau)\}\sin\{\omega_a(t-\tau)\}\,d\tau \tag{7}$$

When the parameters ν and ω are fixed, the integral in Eqn (7) is the well-known Duhamel integral. Differentiation of this equation with respect to time yields an expression for the history of the velocity response:

$$\dot{x}(t) = -\int_0^t a(\tau)\exp\{-\nu\omega(t-\tau)\}\cos\{\omega_a(t-\tau)\}\,d\tau - \nu\omega x(t) \tag{8}$$

Differentiating again gives the absolute acceleration of the oscillator:

$$\ddot{x}(t)+a(t) = \omega_a\int_0^t a(\tau)\exp\{-\nu\omega(t-\tau)\}\sin\{\omega_a(t-\tau)\}\,d\tau - 2\nu\omega\dot{x}(t)-(\nu\omega)^2x(t) \tag{9}$$

The displacement, velocity and acceleration response spectra are defined as functions of ω for fixed damping ratio ν as follows:

$$S_d(\omega;\nu) = |x(t)|_{max}$$

$$= \left|-\frac{1}{\omega_a}\int_0^t a(\tau)\exp\{-\nu\omega(t-\tau)\}\sin\{\omega_a(t-\tau)\}\,d\tau\right|_{max} \tag{10}$$

$$S_v(\omega;\nu) = |\dot{x}(t)|_{max}$$

$$= \left|-\int_0^t a(\tau)\exp\{-\nu\omega(t-\tau)\}\cos\{\omega_a(t-\tau)\}\,d\tau + \nu\omega x(t)\right|_{max} \tag{11}$$

$$S_a(\omega;\nu) = |\ddot{x}(t)+a(t)|_{max}$$

$$= \left|\omega_a\int_0^t a(\tau)\exp\{-\nu\omega(t-\tau)\}\sin\{\omega_a(t-\tau)\}\,d\tau - 2\nu\omega\dot{x}(t)-(\nu\omega)^2x(t)\right|_{max} \tag{12}$$

It should be noted that S_d and S_v are the maximum values of the relative displacement response and the relative velocity response respectively of the single-degree-of-freedom system and S_a is the maximum value of its absolute acceleration response.

The total energy of the single-degree-of-freedom model is

$$E_t(t) = \tfrac{1}{2}m\{\dot{x}(t)\}^2 + \tfrac{1}{2}k\{x(t)\}^2 \tag{13}$$

where m is the mass of the model and $k = m\omega^2$ is its stiffness. The first term on the right-hand side of Eqn (13) is the kinetic energy and the second is the potential energy. It is well known that the kinetic energy is a maximum when the potential energy is a minimum and vice versa. Therefore the maximum velocity response of the system is related to the maximum total energy through the equation

$$|E_t(t)|_{max} = \tfrac{1}{2}m\{\dot{x}(t_{max})\}^2 \tag{14}$$

where the potential energy of the system is assumed to be zero for t_{max}.

Some approximations are introduced in order to obtain simpler expressions for the seismic response spectra (10), (11) and (12). Since the damping ratio is small in civil engineering applications ($2\% \leqslant \nu \leqslant 20\%$), ω_a is substituted by ω and the second non-integral term on the right-hand side of Eqns (11) and (12) is neglected. Moreover, in the computation of the velocity response spectrum (11) the cosine function is approximated by a sine function. Thus three new quantities, defined as displacement, velocity and acceleration pseudo-response spectra, are introduced (Hudson, 1962):

$$S_d{}^*(\omega; \nu) = \left| -\frac{1}{\omega}\int_0^t a(\tau)\exp\{-\nu\omega(t-\tau)\}\sin\{\omega(t-\tau)\}\,d\tau \right|_{max} \tag{15}$$

$$S_v{}^*(\omega; \nu) = \left| -\int_0^t a(\tau)\exp\{-\nu\omega(t-\tau)\}\sin\{\omega(t-\tau)\}\,d\tau \right|_{max} \tag{16}$$

$$S_a{}^*(\omega; \nu) = \left| \omega\int_0^t a(\tau)\exp\{-\nu\omega(t-\tau)\}\sin\{\omega(t-\tau)\}\,d\tau \right|_{max} \tag{17}$$

These approximations allow $S_v{}^*$ and $S_a{}^*$ to be computed from $S_d{}^*$ by using the relations

$$S_v{}^* = \omega S_d{}^* \tag{18}$$

$$S_a{}^* = \omega^2 S_d{}^* \tag{19}$$

Thus the three pseudo-response spectra can be plotted on the same tri-logarithmic scale (Fig. 1).

Duhamel's integral must be evaluated in order to compute the seismic response spectra and pseudo-response spectra. The computational cost of this evaluation can be significantly reduced if the following transformation of Eqn (10) is used:

$$|x(t)|_{max} = \left| -\frac{1}{\omega_a}\int_0^t a(\tau)\exp\{-\nu\omega(t-\tau)\}\sin\{\omega_a(t-\tau)\}\,d\tau \right|_{max}$$

$$= \left| -\frac{1}{\omega_a}\{A(t)\sin(\omega_a t) - B(t)\cos(\omega_a t)\} \right|_{max} \tag{20}$$

where the exponential is written explicitly and the sine function is expanded using

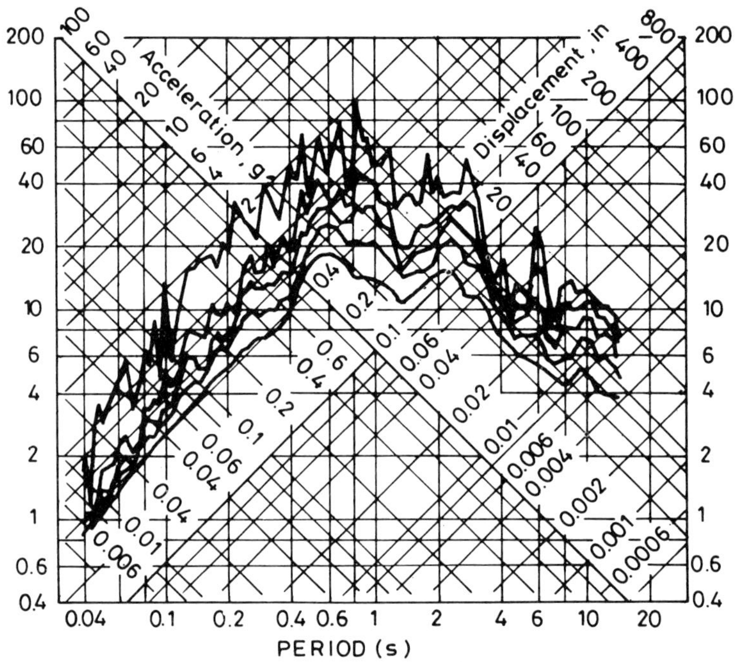

Fig. 1 *Tri-logarithmic representation of a generic response spectrum (from Hudson, 1979).*

the standard trigonometric expression. $A(t)$ and $B(t)$ in Eqn (20) are functions defined by

$$A(t) = \int_0^t a(\tau)\frac{\exp(\nu\omega\tau)}{\exp(\nu\omega t)}\cos(\omega_a\tau)\,d\tau$$

$$B(t) = \int_0^t a(\tau)\frac{\exp(\nu\omega\tau)}{\exp(\nu\omega t)}\sin(\omega_a\tau)\,d\tau \qquad (21)$$

Various numerical procedures, such as the simple summation method, trapezoidal quadrature or Simpson's method, can be used to calculate these integrals efficiently (Clough and Penzien, 1975). A good approximation based on a simple summation method has been proposed by Clough (1970). In this method the dynamic response is computed at equal intervals Δ. Values of the functions (21) are calculated at time t by starting from the values at the previous time $t-\Delta$:

$$A(t) = A(t-\Delta) + \Delta a(t-\Delta)\cos\{\omega_a(t-\Delta)\}$$

$$B(t) = B(t-\Delta) + \Delta a(t-\Delta)\sin\{\omega_a(t-\Delta)\} \qquad (22)$$

The same transformation can be used to evaluate the acceleration response spectrum because the same integrands appear in both Eqn (10) and Eqn (12). The presence of $\cos\{\omega_a(t-\tau)\}$ in the integral of Eqn (11) does not require major numerical changes to be made in the computational procedure for evaluating the velocity response spectrum because the integral can be calculated by performing similar operations to those in expression (20).

Thus to compute the seismic response spectra (10), (11) and (12), the functions $A(t)$ and $B(t)$ given in Eqns (21) only need be evaluated once, and then the cross products with the corresponding trigonometric functions are obtained as shown in Eqn (20). In order to compute the seismic pseudo-response spectra (15), (16) and (17), $S_d{}^*$ is calculated through the evaluation of functions $A(t)$ and $B(t)$. $S_v{}^*$ and $S_a{}^*$ are then obtained using (18) and (19).

Other numerical methods used to compute the seismic response spectra are based on solving the differential equation (6) by direct integration methods (Clough and Penzien, 1975; Bathe, 1982).

3.2 Numerical comparison of the response spectra with the pseudo-response spectra

Two synthetic accelerograms have been generated for comparing the pseudo-response spectra with the response spectra. A plot of each is given in Fig. 2. The

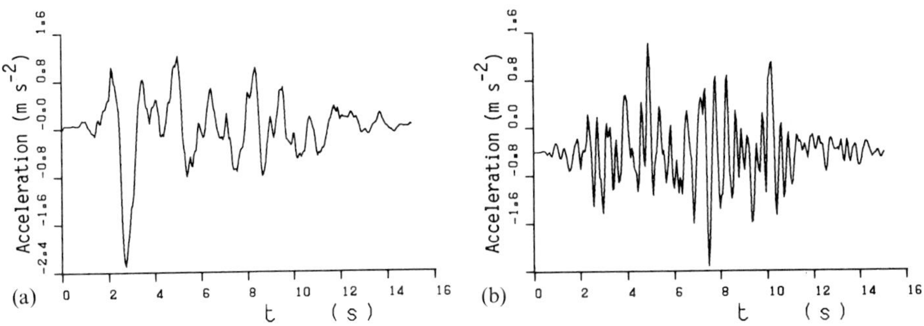

Fig. 2 *(a) Synthetic accelerograms with predominant periods $2\pi/\omega$ of (a) 1.25 s and (b) 0.4 s.*

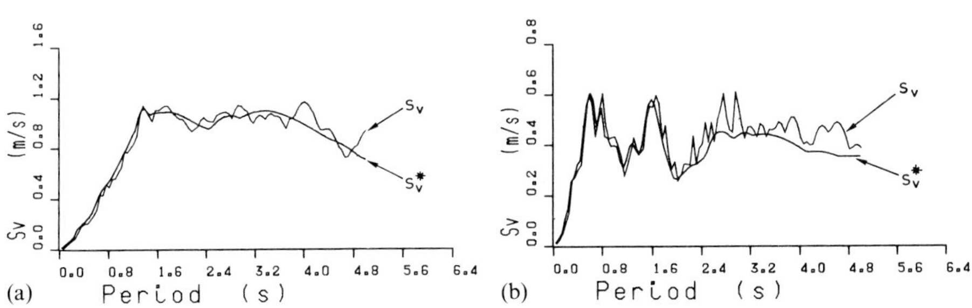

Fig. 3 *Differences between the velocity response spectrum and the velocity pseudo-spectrum: (a) synthetic input accelerogram with predominant period $2\pi/\omega = 1.25$ s; (b) synthetic input accelerogram with predominant period $2\pi/\omega = 0.4$ s.*

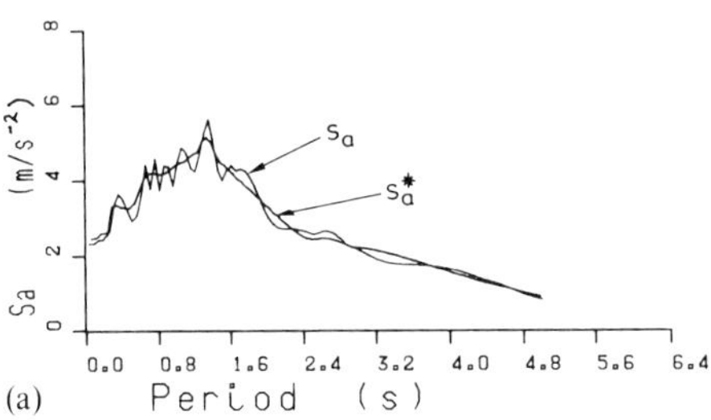

(a)

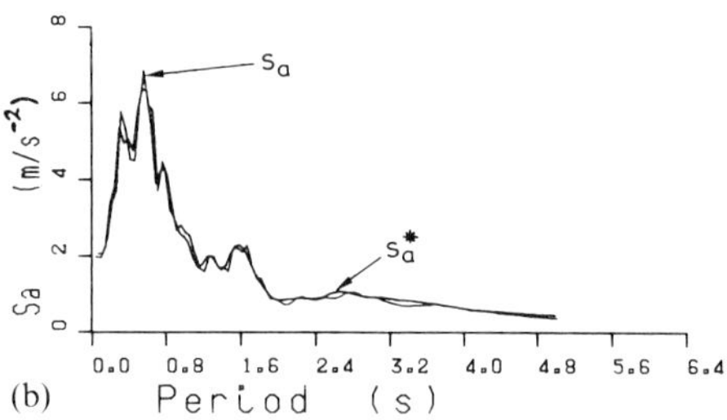

(b)

Fig. 4 Differences between the acceleration response spectrum and the acceleration pseudo-spectrum: (a) synthetic input accelerogram with predominant period $2\pi/\omega = 1.25$ s; (b) synthetic input accelerogram with predominant period $2\pi/\omega = 0.4$ s.

response spectra and the pseudo-response spectra were evaluated for a damping ratio ν of 0.05 and regular sampled periods of 0.05 s between zero and 5 s.

The difference between the displacement spectrum and the pseudo-spectrum is very small, as is expected from Eqns (10) and (15). The discrepancies between the velocity and acceleration spectra and the corresponding pseudo-spectra are shown in Figs 3 and 4. In these two cases the maximum values coincide and the differences are no greater than 15%.

4 Comparison of the amplitude Fourier spectrum with the seismic velocity response spectrum

4.1 Theoretical comparison

In this section the seismic velocity response spectrum with zero damping ratio ($\nu = 0$ and $\omega_a = \omega$) is compared with the amplitude Fourier spectrum of the input acceleration $a(t)$. By assuming a zero damping ratio ($\nu = 0$) Eqn (11) can be rewritten

$$S_v(\omega; \nu = 0) = \left| - \int_0^t a(\tau) \cos\{\omega(t-\tau)\} \, d\tau \right|_{\max} \tag{23}$$

In civil engineering applications it can be assumed that the signal $a(t)$ (i.e. the time history of the ground motion acceleration) has a finite duration T and that it can be expressed in the following form by a zero mean Fourier series:

$$a(t) = \sum_{k=1}^{\infty} a_k \cos(\theta_k t) + \sum_{k=1}^{\infty} b_k \sin(\theta_k t) \qquad \theta_k = \frac{2\pi k}{T} \tag{24}$$

where

$$a_k = \frac{2}{T} \int_0^T a(\tau) \cos(\theta_k \tau) \, d\tau \tag{25}$$

$$b_k = \frac{2}{T}\int_0^T a(\tau)\sin(\theta_k\tau)\,\mathrm{d}\tau$$

Substituting the Fourier series in $\dot{x}(t)$ and using standard trigonometric relations gives a new expression for $\dot{x}(t)$:

$$\dot{x}(t) = -\sum_{k=1}^{\infty}\int_0^t [\cos(\omega t)\{a_k\cos(\theta_k\tau)\cos(\omega\tau) + b_k\sin(\theta_k\tau)\cos(\omega\tau)\}$$
$$+\sin(\omega t)\{a_k\cos(\theta_k\tau)\sin(\omega\tau) + b_k\sin(\theta_k\tau)\sin(\omega\tau)\}]\,\mathrm{d}\tau$$

By applying standard trigonometric expressions, this equation can be rewritten as

$$\dot{x}(t) = -\sum_{k=1}^{\infty}\left\{\cos(\omega t)\left(a_k\int_0^t \frac{1}{2}[\cos\{(\theta_k-\omega)\tau\} + \cos\{(\theta_k+\omega)\tau\}]\,\mathrm{d}\tau\right.\right.$$

$$\left. + b_k\int_0^t \frac{1}{2}[\sin\{(\theta_k-\omega)\tau\} + \sin\{(\theta_k+\omega)\tau\}]\,\mathrm{d}\tau\right)$$

$$+\sin(\omega t)\left(a_k\int_0^t \frac{1}{2}[-\sin\{(\theta_k-\omega)\tau\} + \sin\{(\theta_k+\omega)\tau\}]\,\mathrm{d}\tau\right.$$

$$\left.\left. + b_k\int_0^t \frac{1}{2}[\cos\{(\theta_k-\omega)\tau\} - \cos\{(\theta_k+\omega)\tau\}]\,\mathrm{d}\tau\right)\right\} \tag{26}$$

The integrals in (26) are evaluated for two different cases: firstly for $\theta_k \neq \omega$ and secondly for $\theta_k = \omega$. All terms with $\theta_k \neq \omega$ are bounded in t because the integrals are sine and cosine functions. The terms with $\theta_k = \omega$ in the second and fourth integrals are also bounded in t for the same reason. However, when $\theta_k = \omega$, the first and third integrals are unbounded in t. It can be shown that these integrals have two components: one of them is a sine–cosine function of t and the other is linear in t. Since the systems considered in seismic engineering applications have small natural periods compared with the duration T of the seismic signal, all the bounded terms can be dropped. Thus only the terms with $\theta_k = \omega$ are important and Eqn (26) can be approximated by

$$\dot{x}(t) \approx -\frac{1}{2}t\{a\cos(\omega t) + b\sin(\omega t)\} \tag{27}$$

where a and b are the Fourier coefficients associated with the term with $\theta_k = \omega$. In order to compute the maximum velocity response, the maximum of expression (27) has to be evaluated. The function

$$f(t) = a\cos(\omega t) + b\sin(\omega t) \tag{28}$$

and its first derivative

$$\dot{f}(t) = -a\omega \sin(\omega t) + b\omega \cos(\omega t) \tag{29}$$

are considered. The extrema of function (28) can be obtained by equating the first derivative (29) to zero:

$$-a\omega \sin(\omega t) + b\omega \cos(\omega t) = 0$$

Then its relative extremum occurs at

$$\frac{1}{\omega}\arctan\left(\frac{b}{a}\right) \tag{30}$$

In earthquake engineering applications intermediate and high frequencies are of interest. If the duration of the seismic event is T (between about 20 s and 200 s for strong ground motions), the assumption $2\pi/\omega \ll T$ holds for intermediate and high frequencies (approximately $0.2\,\text{s} < 2\pi/\omega < 2\,\text{s}$). Therefore, from (27), the absolute extremum of $\dot{x}(t)$ will be reached very close to the end of the seismic input (Fig. 5) and

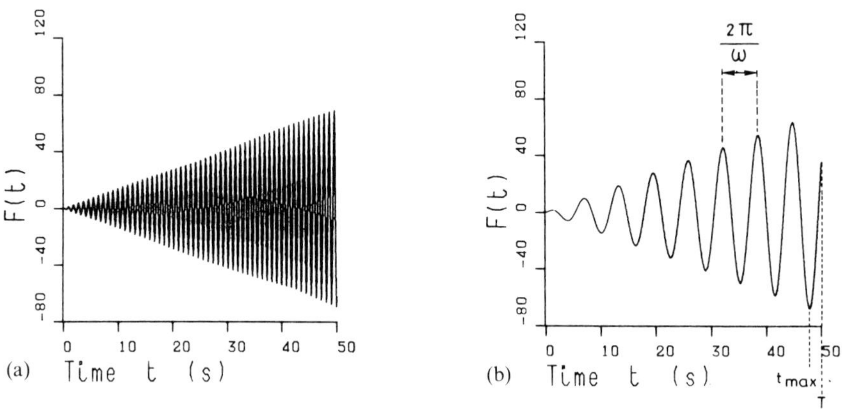

Fig. 5 *Graphical representation of the function* $\frac{1}{2}t\{a\cos(\omega t) + b\sin(\omega t)\}$ *for* $T = 50$ *s and* $a = b = 2$: *(a)* $\omega = 10$; *(b)* $\omega = 0.5$.

$|\dot{x}(t)|_{max} = |\dot{x}(t_{max})|$, $T \approx t_{max}$. Thus the absolute maximum of $\dot{x}(t)$ is approximately given by

$$|\dot{x}(t)|_{max} = |\dot{x}(t_{max})|$$

$$\approx \frac{1}{2}T\left[a\cos\left\{\arctan\left(\frac{b}{a}\right)\right\} + b\sin\left\{\arctan\left(\frac{b}{a}\right)\right\}\right]$$

$$= \tfrac{1}{2}T(a^2+b^2)^{1/2}$$

and the velocity response spectrum can be obtained from

$$|\dot{x}(t)|_{\max} \approx -\tfrac{1}{2}T(a^2+b^2)^{1/2}$$

$$= \left[\left\{\int_0^t a(\tau)\cos(\omega\tau)\,\mathrm{d}\tau\right\}^2 + \left\{\int_0^t a(\tau)\sin(\omega\tau)\,\mathrm{d}\tau\right\}^2\right]^{1/2}$$

$$= F(\omega) \tag{31}$$

This expression is equivalent to the amplitude Fourier spectrum given in (2). Equation (31) shows that, for intermediate and high frequencies, the acceleration amplitude Fourier spectrum is a good approximation of the velocity response spectrum for zero damping ratio.

4.2 Numerical comparison

The only accelerogram recorded to date in Spain is used to compare the results obtained for the amplitude Fourier spectrum with the velocity response spectrum with zero damping ratio. The longitudinal and transverse components of the accelerogram are shown in Fig. 6.

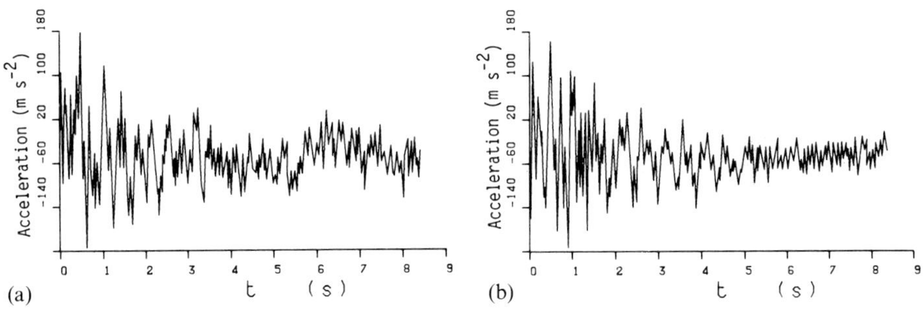

Fig. 6 *Accelerogram obtained from the Beznar Dam, 24 June 1984: (a) longitudinal component; (b) transverse component.*

As has been previously discussed, the approximation of the velocity response spectrum with a zero damping ratio (Eqn (31)) by the amplitude Fourier spectrum is valid only if the predominant period of the seismic excitation signal is small compared with the duration of the seismic signal. The velocity response spectrum S_v (Eqn (11)) and the Fourier spectrum $|F(\theta)|$ (Eqn (2)) shown in Fig. 7 are in good agreement. Figure 8 shows a comparison of the velocity response $\dot{x}(t)$ obtained for the accelerogram with the approximate response given by Eqn (27) for $\omega = 26.889\ \mathrm{rad\,s^{-1}}$. Although these two velocity time histories are different, their maximum values are in good agreement. The maxima are reached at times close to

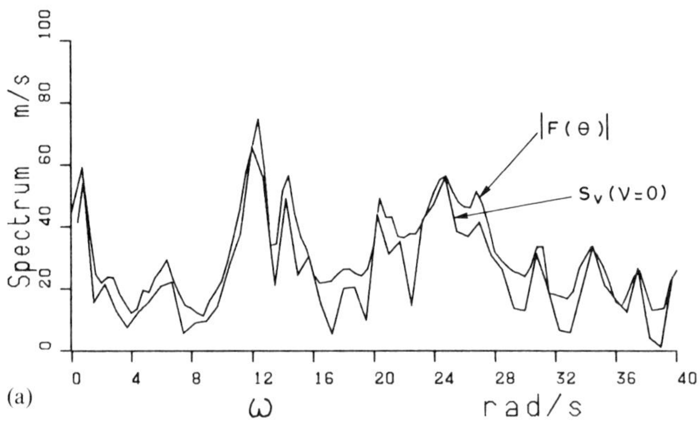

(a)

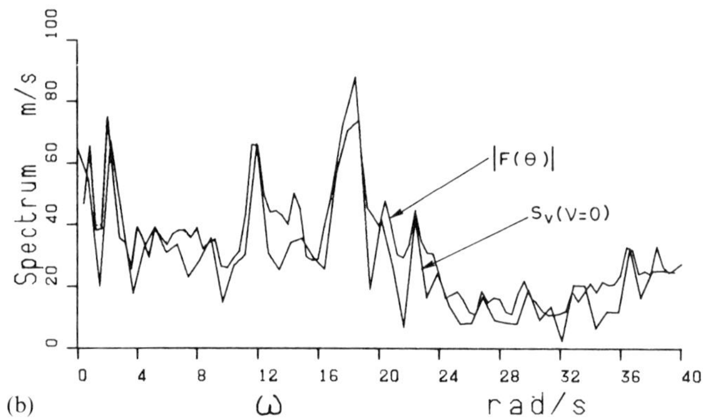

(b)

Fig. 7 *Numerical comparison of the amplitude Fourier spectrum, the velocity response spectrum with zero damping ratio and the velocity pseudo-response spectrum: (a) longitudinal component; (b) transverse component.*

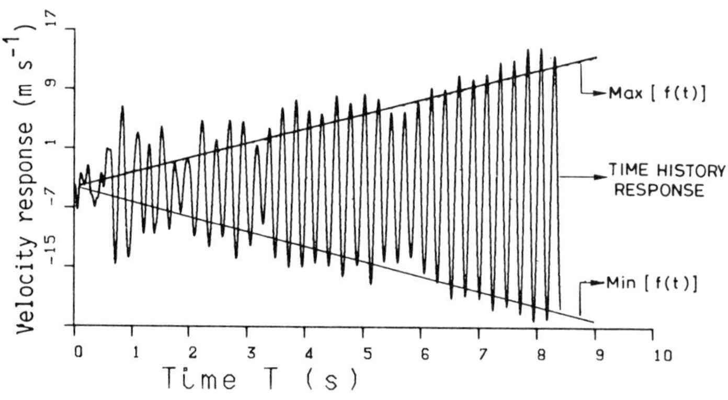

Fig. 8 *Numerical comparison of the time history of the velocity response for the longitudinal component of the Beznar accelerogram with the maximum and minimum of the function given by Eqn (27).*

the total duration of the excitation. It can be seen from Eqns (14) and (31) that

$$E_t(t)|_{max} = \tfrac{1}{2}m\{\dot{x}(t)\}_{max}{}^2 \approx \tfrac{1}{2}m|F(\omega)|^2 \quad (32)$$

5 Concluding remarks

A theoretical study has been performed which demonstrates that the seismic velocity response spectrum with zero damping represents a good approximation of the Fourier amplitude spectrum. An analytical approach was used and the limitations of the approximation were pointed out. These limitations are as follows: the signal is assumed to behave as a Fourier series; the natural period of the system has to be small compared with the duration of the seismic signal. Both conditions are fulfilled in the case of earthquake engineering applications, and thus the approximation is valid for seismic excitations. This approach has been applied to a practical example (the only available accelerogram in Spain to our knowledge) and it has been confirmed that the approximation works reasonably well for high and intermediate frequencies.

Acknowledgment

This research was supported in part by the CAICYT (project 1151/84).

References

BATHE K. J. (1982): *Finite Element Procedures in Engineering Analysis*, Prentice-Hall, Englewood Cliffs, NJ.

BRIGHAM E. O. (1974): *The Fast Fourier Transform*, Prentice-Hall, Englewood Cliffs, NJ.

CLOUGH R. W. (1970): Earthquake response of structures. In WIEGEL R. L. (ed.), *Earthquake Engineering*, pp. 307–334, Prentice-Hall, Englewood Cliffs, NJ.

CLOUGH R. W. and PENZIEN J. (1975): *Dynamics of Structures*, pp. 75–92, McGraw-Hill, New York.

COOLEY J. W., LEWIS P. A. M. and WELCH P. D. (1967): Historical notes on the fast Fourier transform, *IEEE Transactions on Audio and Electronics*, **15** (2), 76–79.

DUARTE R. T. (1987): Earthquake vibration modelling using gaussian processes. In KORIDZE A. (ed.), *Engineering Aspects of Earthquake Phenomena*, pp. 45–57, Omega Scientific, Wallingford.

HOUSNER G. W. (1970): Strong ground motion. In WIEGEL R. L. (ed.), *Earthquake Engineering*, pp. 75–92, Prentice-Hall, Englewood Cliffs, NJ.

HUDSON D. (1962): Some problems in the application of spectrum techniques to strong motion earthquake analysis, *Bulletin of the Seismological Society of America*, **52**, 417–430.

HUDSON D. (1979): *Reading and Interpreting Strong Motion Accelerograms*, Engineering Monographs on Earthquake Criteria, Structural Design, and Strong Motion Records, Earthquake Engineering Research Institute, Berkeley, CA.

SINGLETON R. C. (1969): An algorithm for computing the mixed radix fast Fourier transform, *IEEE Transactions on Audio and Electronics*, **17** (2), 93–103.

TRETTER S. A. (1976): *Introduction to Discrete-time Signal Processing*, Wiley, New York.

On Short-term Earthquake Prediction

GIUSEPPE GRANDORI

Politecnico di Milano, Piazza Leonardo da Vinci 32, I-20133 Milano, Italy

Abstract A statistical analysis of the foreshock–main shock correlation for some seismically active regions in Italy is presented. It is found that the probability that a weak shock will be followed within 2 days by a main shock is of the order of 2%, while the probability that a main shock will be preceded by a foreshock is of the order of 50%. The effectiveness of alarm systems based on a pair of short-term earthquake precursors is analysed.

1 Introduction

A number of papers dealing with short-term earthquake precursors consisting of weak seismic shocks which are regarded as potential foreshocks have appeared recently (Bakun and Lindh, 1985; Jones, 1985; Kagan and Knopoff, 1987). The results reported in these papers together with results obtained by Grandori *et al.* (1988) lead to the following preliminary conclusions.

(1) The characteristics of short-term earthquake precursors consisting of potential foreshocks are almost the same in different regions. The probability that a weak shock will be followed within a few days by a main shock is of the order of 2%, whereas the probability that a main shock is preceded by a foreshock is of the order of 50%.

(2) The combination of foreshocks with another precursor, independent of the foreshocks and with similar characteristics, could provide a reasonably effective alarm system.

The search for a second precursor faces a major difficulty: the collection of statistical data about false and missed alarms in a given region requires a very long observation period when interest is focused on precursors of potentially destructive earthquakes. However, the fact that foreshocks in different regions show almost the same characteristics offers some hope for future research. In fact, if another precursor with constant characteristics over different regions is identified, the required statistical data could be collected by using a large spatial window instead of a long time window in a limited area. This approach would obviously demand international cooperation.

The results reported in detail by Grandori *et al.* (1988) and Jones (1985) are summarized in this paper. These results clarify the conditions under which the precursor consisting of potential foreshocks could be combined with another precursor to provide a reasonably effective alarm system.

2 Definitions

The main parameters involved in short-term earthquake prediction are defined following Grandori *et al.* (1988). Let F be the precursor, E the potentially damaging earthquake and Δt the alarm lifetime. If F occurs but is not followed by E in the time interval Δt, F is denoted a false alarm. The correlation between F and E is defined using the following probabilities: the false alarm probability

$$p = P(\bar{\mathrm{E}}/\mathrm{F}) \tag{1}$$

i.e. the probability that, given F, there will be no E within the interval Δt; the missed alarm probability

$$q = P(\bar{\mathrm{F}}/\mathrm{E}) \tag{2}$$

i.e. the probability that in the interval Δt immediately before E there has been no F; and the background probability $P(\mathrm{E})$ of the event E, i.e. the probability that E will occur in the next interval Δt which is evaluated at a given time without taking any precursors into account. $P(\mathrm{E})$ is actually a conditional probability, given all the information provided by the knowledge of the occurrence process at the time at which the prediction is performed. The probabilities p and q depend on the characteristics of the coupled-point process of events E and F. In general the values of p and q are not constant in time but depend on the history of the stochastic process. If Δt, p, q and $P(\mathrm{E})$ are known, all the probabilities involved in decision-making processes can be determined. In particular, the probability that if F has occurred there will be an event E within Δt is obviously

$$P(\mathrm{E}/\mathrm{F}) = 1 - p \tag{3}$$

For completeness it is also necessary to allow for the fact that, if $q > 0$, unpredicted earthquakes will occur.

A quantity providing a concise measure of the effectiveness of the alarm system, allowing for both p and q, could be defined as follows:

$$U = \frac{s}{f+m} \tag{4}$$

where s is the expected number of successes (alarms followed by an event E within Δt) over a long period of time in which the system is operating, and f and m are the expected numbers of false and missed alarms over the same period of time. The effectiveness U defined by (4) is thus substantially the ratio of the expected number of successful predictions to the expected number of failures of the system.

Equation (4) can be suitably modified whenever social, economic or psychological factors require the attribution of different relative weights to the successes, the false alarms and the missed alarms. If Eqn (4) is accepted, the effectiveness U in stationary conditions can be calculated as a function of p and q (Grandori *et al.*, 1984):

$$U = \frac{(1-p)(1-q)}{p(1-q)+q(1-p)} \tag{5}$$

3 *Foreshock–main shock correlation*

Analysis of the statistics for Southern California (Jones, 1985) shows that the probability that a weak shock will be followed within 5 days by a main shock with

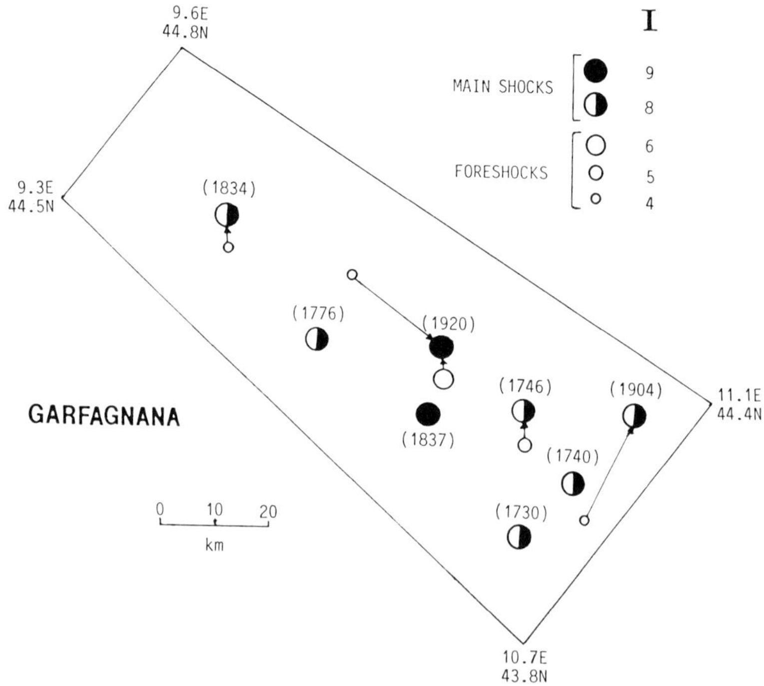

Fig. 1

$M \geqslant 5.0$ increases with the magnitude of the weak shock from less than 1% for $M = 3$ to 5% for $M = 4.5$.

A statistical analysis has been carried out for the Garfagnana region in Italy (Grandori *et al.*, 1988). Seismic shocks with intensities between 4 and 6 on the MCS scale were considered as potential foreshocks (Fig. 1). Potentially damaging earthquakes were taken to have $I(\text{MCS}) \geqslant 8$. An alarm lifetime Δt of 2 days was assumed, and a maximum foreshock–main shock distance of 30 km was adopted for the space window.

The results obtained by Grandori *et al.* are shown in Table 1. The first three rows of the table contain data for weak shocks with $I = 4$, $I = 5$ or $I = 6$ respectively. In the fourth row the data are given for the case when any shock with $4 \leqslant I \leqslant 6$ is considered as a potential foreshock. Table 1 shows that the false alarm probability decreases as the intensity I of the potential foreshock increases. However, the missed alarm probability also increases with I. The most attractive precursor is obtained for the class $4 \leqslant I \leqslant 6$, for which the false alarm probability is almost the same as for the other classes but the missed alarm probability is much smaller.

Table 1 *Statistical analysis of data for the Garfagnana region (n(E) = 8)*

I	n(F)	n(F ∩ E)	p	q
4	137	3	0.978	0.62
5	81	2	0.975	0.75
6	27	1	0.963	0.87
$4 \leqslant I \leqslant 6$	210	4	0.981	0.50

These results are similar to those obtained by Jones (1985) for Southern California. In addition, the results of analyses performed for two other Italian seismic zones (Friuli, northeastern Italy, and Irpinia, southern Italy) are almost the same as those from Garfagnana. The averages of the results obtained from the three Italian regions are presented in Table 2. The short-term precursor consisting of potential foreshocks appears to have almost the same characteristics in many different regions, i.e. a high false alarm probability ($p \approx 0.98$) and a relatively low missed alarm probability ($q \approx 0.5$).

Table 2 *Average results from Garfagnana, Friuli and Irpinia (n(E) = 26)*

I	n(F)	n(F ∩ E)	p	q
4	341	8	0.977	0.69
5	187	5	0.973	0.81
6	77	3	0.961	0.88
$4 \leqslant I \leqslant 6$	530	12	0.977	0.54

The catalogue can be considered complete as far as E events are concerned. However, the data for F and consequently $F \cap E$ events may be incomplete. If this incompleteness were uniformly distributed throughout the catalogue, i.e. if the proportion of missing F events were the same in all subperiods of the period under consideration, however small, the numbers $n(F)$ and $n(F \cap E)$ would be increased proportionally. Thus p would remain unchanged whereas q would decrease. Even with the extreme assumption that all E events have actually been preceded by a precursor F and that, conversely, no false precursor has been lost along the way, the false alarm probability would decrease only slightly. For example, $p = 0.963$ would be obtained for the Garfagnana region. However, this assumption is unrealistic as it would lead to $q = 0$. In conclusion, incompleteness of the catalogue would not substantially alter the value of p and only a minor decrease in q would be expected.

4 *Double-precursor systems*

Consider two independent precursors F_1 and F_2. It should be emphasized that the independence of the two *precursors* does not mean that the two *events* F_1 and F_2 are statistically independent. This would contradict the fact that both precursors are related to event E. We say that the two precursors are independent if events F_1 and F_2 are correlated with each other *only* through event E, with no other direct correlation.

If a state of alarm is proclaimed only when the two warning signals overlap, the false alarm probability $p^{(2)}$ and the missed alarm probability $q^{(2)}$ of the system are given by (Grandori *et al.*, 1984)

$$p^{(2)} = 1 - \left\{ 1 + \frac{P(E)}{1 - P(E)} \frac{p_1}{1 - p_1} \frac{p_2}{1 - p_2} \right\}^{-1} \tag{6}$$

$$q^{(2)} = 1 - (1 - q_1)(1 - q_2) \tag{7}$$

A precursor is classified as "weak" either if both p and q are very high or if at least one of them is very high. Only weak precursors are considered here because, obviously, if a strong precursor were available (i.e. a precursor with both p and q small) the problem would be satisfactorily solved by this precursor alone.

In Table 3 the results obtained for single precursors are compared with those obtained for pairs of independent precursors. Three types of precursor are considered: (a) both p and q are very high; (b) q is very high and p is relatively low; (c) p is very high and q is relatively low. The characteristics of alarm systems based on pairs of independent precursors which are both of type (a), both of type (b) or both of type (c) are given. The value of $P(E)$ in Eqn (8) is assumed to be 0.000 25. In stationary conditions this value corresponds to a mean time $T(E)$ between main shocks of 22 years with an alarm lifetime Δt of 2 days.

The false alarm probability for the double-precursor system is always extremely small compared with that for the single precursor. However, the missed alarm probability is larger than that for the single precursor. In particular, in cases (a) and (b) only two in a thousand main shocks would be predicted. The influence of this fact on the effectiveness of the double-precursor system is such that $U^{(2)}$ is 10 times smaller than U in case (a) and 20 times smaller in case (b). In contrast, two precursors of type (c) provide a much more effective alarm system than the single precursor.

Table 3 *Comparison of single- and double-precursor systems*

Type	*Single precursor*			*Double precursor*		
	p	q	U	$p^{(2)}$	$q^{(2)}$	$U^{(2)}$
(a)	0.96	0.96	0.02	0.125	0.998	0.002
(b)	0.4	0.96	0.04	0.0001	0.998	0.002
(c)	0.96	0.4	0.04	0.125	0.64	0.52

In conclusion, advantages are only achieved by combining weak precursors of type (c). In other words, an alarm system based on a pair of independent precursors appears to have very interesting properties, even if their individual false alarm probabilities are very high, provided that the missed alarm probability of each precursor is relatively low. As pointed out in the preceding section, potential foreshocks constitute a precursor that does in fact have such properties. Unfortunately, a second precursor which is independent of foreshocks and has similar properties has not yet been identified. However, the discussion presented above indicates a possible approach to the development of an effective alarm system based on a pair of weak precursors. It is of crucial importance that the two precursors are independent. As shown by Grandori *et al.* (1988), even a small degree of correlation between precursors will drastically reduce the advantages of the double-precursor system with respect to the false alarm probability.

References

BAKUN W. H. and LINDH A. G. (1985): The Parkfield, California, prediction experiment, *Earthquake Prediction Research*, **3**, 285–304.

GRANDORI G., GUAGENTI E. and PEROTTI F. (1984): Some observations on the probabilistic interpretation of short-term earthquake precursors, *Earthquake Engineering and Structural Dynamics*, **12**, 749–760.

GRANDORI G., GUAGENTI E. and PEROTTI F. (1988): Alarm systems based on a pair of short-term earthquake precursors, *Bulletin of the Seismological Society of America*, to be published.

JONES L. (1985): Foreshocks and time-dependent earthquake hazard assessment in Southern California, *Bulletin of the Seismological Society of America*, **75**, 1669–1679.

KAGAN Y. and KNOPOFF L. (1987): Statistical short-term earthquake prediction, *Science*, **236**, 1563–1567.

Bibliography

AKI K. (1981): A probabilistic synthesis of precursory phenomena. In *Earthquake Prediction: An International Review*, pp. 566–574. American Geophysical Union, Washington, DC.

GRANDORI E. and GRANDORI G. (1983): Some probabilistic aspects of earthquake prediction, *Meccanica, Journal of the Italian Association of Theoretical and Applied Mechanics*, **18**, 246–255.

UTSU T. (1979): Calculation of the probability of success of an earthquake prediction (in the case of Izu–Oshima–Kinkai earthquake of 1978), *Reports of the Coordinating Committee for Earthquake Prediction*, **21**, 1964–1966 (in Japanese).

International Cooperation in the Field of Earthquake Engineering

A. KORIDZE

UNESCO, Paris, France

The surface of our planet is exposed to a variety of natural and man-made hazards. Earthquakes, which are among the most severe of natural hazards, cause large human and economic losses in various parts of the world. During the last two decades a number of destructive earthquakes have claimed thousands of human lives and caused enormous material losses in Latin America, Asia, the Balkans and the Arab States. The earthquake which occurred in Tangshan City, China, on 28 July 1976 was the worst natural disaster to strike modern civilization. It killed at least 240 000 people and seriously injured 160 000 more. Many unreinforced masonry buildings collapsed, and this was the main cause of the huge loss of human life. The problem of earthquakes becomes even more complicated for developing countries where, owing to the combination of several factors (lack of codes and regulations, low quality of construction etc.), the scale of material damage is frequently far beyond the economic capabilities of the affected region, thus causing unforeseen social and economic disruption. However, earthquake effects on similar types of structures built from similar materials are almost identical throughout the world, although with certain differences depending on the quality of construction and local seismotectonic conditions. Therefore the need for intensive international cooperation in the field of earthquake engineering is quite evident.

The UNESCO-inspired international cooperation in the field of earthquake engineering started in the early 1960s when UNESCO undertook responsibility as the executive agency of the United Nations Development Programme (UNDP) for several large-scale projects in training and research in the various regions of the world. Since then, for a period of about 25 years, UNESCO, in close cooperation with other United Nations bodies, including the Office of the United Nations Disaster Relief Coordinator (UNDRO) and the United Nations Industrial Development Organization (UNIDO), has implemented a number of important projects at national and regional level and organized seminars and training courses, study missions and post-earthquake observation missions. However, during the execution of a number of projects it was recognized that there was a lack of an international mechanism for the exchange of experience and information in the field of earthquake engineering on a permanent basis, so that it was not possible to transfer knowledge from advanced scientific centres to concerned centres in developing countries. This gap of knowledge in earthquake engineering has been increasing from year to year.

The idea of promoting this kind of exchange of information systematically was nurtured by the participants of the workshop for the establishment of the UNESCO International Network of Earthquake Engineering Centres (INEEC). This meeting was organized by the International Centre for Mechanical Sciences (CISM) in Udine, Italy, in April 1986. Representatives from well-known centres specializing in earthquake engineering in Italy, Yugoslavia, Turkey and the USSR took part in the meeting. It was agreed that INEEC would be an international network of autonomous national educational or research centres covering, under the auspices of UNESCO, the following fields of earthquake engineering: seismic codes, seismic risk analysis, vulnerability and strengthening criteria, post-earthquake damage evaluation, low cost housing in seismic areas, response of structures and cost–benefit analysis in earthquake engineering. Today, the following institutions are among the focal points of INEEC: International Centre for Mechanical Sciences, Udine, Italy; National Laboratory for Civil Engineering, Lisbon, Portugal; Boğaziçi University, Istanbul, Turkey; Academy of Sciences of the Georgian SSR, Tbilisi, USSR; Institute of Earthquake Engineering and Engineering Seismology, University "Kiril i Metodij", Skopje, Yugoslavia.

UNESCO, INEEC and other United Nations agencies cannot cover all international cooperation in this field since there are many forms of cooperation established through bilateral or multilateral programmes. One of the main aims of such cooperation is the transfer of knowledge on earthquake engineering, testing methods and software, and appropriate testing equipment for earthquake engineering laboratories. This also means that collaborative projects should be set up for the study of the most important problems of earthquake engineering in developing countries, involving scientific centres, members of INEEC and other interested national institutions and international organizations. In view of the problems and primary needs of developing countries concerning seismic risk mitigation, these projects should be organized in the following areas:

(1) the classification and mapping of urban and rural dwellings;

(2) performance evaluation and vulnerability studies;

(3) the development of national codes and standards for aseismic design and construction;

(4) the development of appropriate low cost testing facilities for materials studies, models and full-scale structure testing;

(5) the creation of databanks for different aspects of performance, vulnerability, design, repair and reconstruction of low cost housing structures.

It is most important to establish cooperation in the field of development of compatible methodologies for the evaluation of earthquake damage and economic losses which could be applied widely in a region or several countries. This would provide the basis for comparative studies on the damageability of similar structural building types during different earthquakes in different countries and allow a re-evaluation of the consequences of past earthquakes.

The education and training of professionals is of primary importance for creating conditions to undertake long-term measures and the development of a national policy for seismic risk mitigation. A country lacking national professional staff cannot be expected to be able to deal successfully with earthquake damage and loss-reduction problems. Therefore another important goal of international cooperation is to assist the efforts of developing countries in the creation of national professional staff by training them in scientific or educational institutions with a high level of experience in earthquake engineering. Among the most outstanding international centres in training established by the initiative and financial support of UNESCO are the Institute of Earthquake Engineering and Engineering Seismology (IZIIS), Skopje, Yugoslavia, and the International Institute of Seismology and Earthquake Engineering (IISEE), Tokyo, Japan. Since the creation of these centres, international experts have been transferring their knowledge and experience in teaching and research in engineering seismology and earthquake engineering in their respective countries. About 100 students from developing countries have been awarded fellowships by UNESCO, UNDP and other international organizations, thus enabling them to participate in the courses organized in these institutions.

In most developing countries located in seismically hazardous areas, earthquake engineering as a scientific discipline is not included in university, post-graduate or advanced training programmes. However, several countries have recently launched projects or expressed their wish to create such programmes in cooperation with UNESCO and INEEC. Among them is Algeria, where a national centre for applied research in earthquake engineering is in the process of being created and will play a regional and international role in the future. During the international seminars for Maghrebian countries on the seismic resistance of low cost housing structures held in Casablanca, Morocco, in 1986 and in Tunis, Tunisia, in 1987, it was decided to organize, in cooperation with INEEC and the UNESCO International Technical University (ITU), a postgraduate course in earthquake engineering for specialists from these countries and later to put this course on video and distribute it in video cassette form for distance training in this region.

In general, training and education programmes for developing countries should consist of the following:

(1) the training of engineers for the degree of master or doctor of technical sciences in the field of earthquake engineering and engineering seismology;

(2) the organization of long-term courses (3–6 months) in some of the centres for the training of engineers or junior scientific staff for the needs of developing countries or countries with less developed earthquake engineering;

(3) the organization of short-term courses (2–3 weeks) for the training of local

specialists (architects, engineers, technicians), skilled workers and self-builders in the basic principles of aseismic requirements for construction of low cost housing.

Another very important activity is related to the creation of mechanisms for the transfer and dissemination of knowledge and the provision of scientific information. These objectives could be accomplished through the following:

(1) the organization of specialized scientific meetings on the most important problems of earthquake engineering, such as vulnerability reduction in low cost urban and rural buildings (it is desirable for participants to be selected from countries with similar climates, environments, local building materials and traditional construction features);

(2) providing conditions for transferring software and laboratory equipment, particularly appropriate low cost equipment, from developed to developing countries or to countries which lack knowledge in earthquake engineering;

(3) publication of volumes including papers of eminent scientists, bulletins and reviews of results of research work in various national institutions, review reports on various seminars and the implementation of appropriate new findings in the development of preventive measures in the construction of low cost housing;

(4) the installation of an electronic mailing system, at least in the focal-point institutions of international networks.

In conclusion, it can be said that, owing to the nature of earthquakes which respect no political frontiers, the development and strengthening of international cooperation in the field of earthquake engineering will reduce the gap in knowledge between developed and developing countries and solve a number of important problems of seismic risk reduction at lower cost and in a relatively short time.